M000222568

MANAGING YOUR DATA SCIENCE PROJECTS

LEARN SALESMANSHIP, PRESENTATION, AND MAINTENANCE OF COMPLETED MODELS

Robert de Graaf

Apress®

Managing Your Data Science Projects: Learn Salesmanship, Presentation, and Maintenance of Completed Models

Robert de Graaf
Kingsville, VIC, Australia

ISBN-13 (pbk): 978-1-4842-4906-2
https://doi.org/10.1007/978-1-4842-4907-9

ISBN-13 (electronic): 978-1-4842-4907-9

Copyright © 2019 by Robert de Graaf

This work is subject to copyright. All rights are reserved by the Publisher, whether the whole or part of the material is concerned, specifically the rights of translation, reprinting, reuse of illustrations, recitation, broadcasting, reproduction on microfilms or in any other physical way, and transmission or information storage and retrieval, electronic adaptation, computer software, or by similar or dissimilar methodology now known or hereafter developed.

Trademarked names, logos, and images may appear in this book. Rather than use a trademark symbol with every occurrence of a trademarked name, logo, or image we use the names, logos, and images only in an editorial fashion and to the benefit of the trademark owner, with no intention of infringement of the trademark.

The use in this publication of trade names, trademarks, service marks, and similar terms, even if they are not identified as such, is not to be taken as an expression of opinion as to whether or not they are subject to proprietary rights.

While the advice and information in this book are believed to be true and accurate at the date of publication, neither the authors nor the editors nor the publisher can accept any legal responsibility for any errors or omissions that may be made. The publisher makes no warranty, express or implied, with respect to the material contained herein.

Managing Director, Apress Media LLC: Welmoed Spahr
Acquisitions Editor: Shiva Ramachandran
Development Editor: Laura Berendson
Coordinating Editor: Rita Fernando

Cover designed by eStudioCalamar

Author photo © Elleni Toumpas

Distributed to the book trade worldwide by Springer Science+Business Media New York, 233 Spring Street, 6th Floor, New York, NY 10013. Phone 1-800-SPRINGER, fax (201) 348-4505, e-mail orders-ny@springer-sbm.com, or visit www.springeronline.com. Apress Media, LLC is a California LLC and the sole member (owner) is Springer Science + Business Media Finance Inc (SSBM Finance Inc). SSBM Finance Inc is a **Delaware** corporation.

For information on translations, please e-mail rights@apress.com, or visit http://www.apress.com/rights-permissions.

Apress titles may be purchased in bulk for academic, corporate, or promotional use. eBook versions and licenses are also available for most titles. For more information, reference our Print and eBook Bulk Sales web page at http://www.apress.com/bulk-sales.

Any source code or other supplementary material referenced by the author in this book is available to readers on GitHub via the book's product page, located at www.apress.com/9781484249062. For more detailed information, please visit http://www.apress.com/source-code.

Printed on acid-free paper

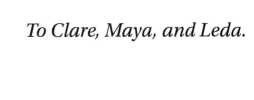

To Clare, Maya, and Leda.

Contents

About the Author

Robert de Graaf is currently a data scientist at RightShip and was central to the development of the algorithm currently used in the Qi platform to predict maritime accidents, among other models. He initially began his career as an engineer, at different times working in quality assurance, project engineering, and design, but soon became interested in applying statistics to business problems and completed his education with a master's degree in statistics. He is passionate about producing data solutions that solve the right problem for the end user.

Introduction

Data scientists, possibly especially in the early stages of their career, are exhorted to learn about a great number of different machine learning techniques and algorithms. We are told to learn about deep learning, support vector machines, and Random Forests, to name only a few.

The truth is that no one really cares about machine learning algorithms. They don't care about Random Forests, they don't care about support vector machines, and they very much don't care about convolutional neural networks or the better and newer deep learning architecture you spent hours deciphering papers and ingesting online lectures to understand.

They just want their problem solved.

It's the same if you get a plumber to unblock your toilet. They might tell you what they're going to do, but you don't really care—you just want a working toilet.

The plumber repairing a broken fixture is usually in a much better position than a data scientist. There is no doubt what the problem is, no doubt about the essential characteristics of a working toilet; the plumber won't need to explain to you why it's useful to have one.

Data science falls down here. Often the client or customer doesn't quite know what the problem is; they just know things could be better in some area of their business. Even if they do know what the problem is, that doesn't have to mean they have a clear idea of what things would look like if that problem was solved.

As a result the first task of a data scientist is to figure out what the problem is, and then to figure out whether a viable solution using data science exists, and if so, what it looks like. In turn, this means that a data scientist, whether tasked with providing solutions for an in-house team or directly selling data science to an external customer, needs to spend at least part of their time thinking and acting like a management consultant.

I'm far from the only person to feel this. Writing in the Harvard Business Review, Kalyan Veeramachaneni of MIT posed the question "Why You're Not

Getting Value From Data Science."[1] One of the first reasons he provides is that not enough attention is given by data scientists to the objective of driving business value; data scientists want to build interesting models instead.

My prediction is that either data scientists will become more focused on the goal of driving business value or there will be far fewer in the future, and many of those who are left will lose their autonomy.

Either way, where the Veeramachaneni article contends that most data scientists want to be primarily machine learning experts, I expect that machine learning experts represent only a small proportion of the data science job market.

This book hopes to promote the first outcome by not just equipping data scientists to be better focused on business value but being able to show others around them that they have that focus. The latter half is very important because value is in the eye of the beholder. Unless the people around you know what you are doing for them and understand how much it improves their business, they won't trust you to deliver in the future.

Thinking like a Consultant

To a degree, every data scientist needs to think like a consultant. This is most obviously true for data scientists who take on projects for other companies but also remains true for data scientists who only ever have customers within the organization they work for. To be successful you need to adopt many of the behaviors of effective consultants.

The reason is that the same loop is needed for identifying where there is value, gaining the customer's trust to have license to work toward that value, delivering and then showing that you delivered, which renews and extends your license to look at their problems into the future.

One of the most significant challenges for a consultant is to gain their client's trust, so that there can be a genuine partnership in the quest to improve a business. In Chapter 4, we will look at the Trust Equation, one way to frame the kind of relationship you need to build with your clients or customers to work effectively, which has been popular among consultants.

However, we'll also extend the logic to the models themselves, as in some way users need to trust models more than they need to trust you—they can't haul the model over the coals and demand answers if something goes wrong.

---

[1]Kaylan Veeramachaneni, "Why You're Not Getting Value From Your Data Science Team," Harvard Business Review, https://hbr.org/2016/12/why-youre-not-getting-value-from-your-data-science.

Most of everything else follows from this crucial fact—users won't use a model they can't trust, and if they don't use your work, your chances of getting to do more work of the kind you enjoy (whatever that is) are diminished.

Data Science Product Life Cycle

Not long ago, I attended a job interview where I was asked how many models I usually made. I don't remember the answer I gave, but I'm sure it was too low, as the interviewer went on to explain that the expectation was for three or four new models every week.

Reading this book and looking at the required activities, it should be clear that that sort of a pace is unlikely if models are to be implemented and solve problems faced by real people by being continually used by people. The approach I'm promoting in this book is for fewer more targeted, more carefully reviewed, more carefully constructed, and more carefully implemented models. Each new model requires its own iterative process to ensure that its benefits are fully realized and fully communicated.

If you are a manager of a larger IT or innovation function or are otherwise a non-data scientist senior manager sitting somewhere above a data science team, this book should give you a picture of the most useful ways that a data science team can be employed.

At the highest level, if you only ever see your data science team writing esoteric code, or comparing small differences in accuracy scores between model iterations, you won't get the best value from them.

Their product isn't models—it's improvements to the way different groups within your business do their jobs. The model is one element of the package your data science team should deliver with each project—by reading this book you should get a picture of how the rest of the package should look.

On with the Show

The core of this book centers on data science projects. This core is sandwiched by an opening chapter on how to strategize for a team and a final chapter on how to get the best out of a team. Projects are the essential unit of composition for data science—what you implement by the end of a project is what is visible to users and customers. However, there is still a need to step back from the projects and think about the big picture for your team, both in terms of where the team is heading and in terms of how it can work best to get there. The first and last chapters provide this perspective.

The core structure of the book is designed to guide you from beginning to end on a data science project. We start with understanding how to set objectives for a data science team, then how to set objectives for individual data science projects travel through, to how to ensure you get the proper credit for your projects. At each stage, we will tie what needs to be done to the business requirements. As a result, people with the full range of backgrounds in technical science can read this book and learn something, from beginner or non-data scientist manager through to machine learning expert.

Regardless of your background, I hope you find something here you can take back to your organization and use to ensure that the data science products that are built there succeed and go on to instill confidence that data science is a great way to build solutions for the problems that are found in many organizations.

Data Science Team Strategy

The Right Game Plan

If you don't plan where you are going from the start, you can't expect to be satisfied with your final destination. In the business world, as in many others, having a well-defined strategy, and working to that strategy, ensures you arrive where you hoped.

In the context of data science, strategy can be applied at a number of different levels. There are distinct contrasts between applying a strategy for the whole organization, for a whole team, and for a particular project.

This chapter takes a high-level view, working through the process of developing a data science team strategy that is aligned to the goals of the organization as a whole, creating a document that records and communicates the strategy, and then considers how to use the strategy to the best advantage of the organization by referring to it for decision making.

A lot of the emphasis in this chapter will be on people who are in data science teams in larger organizations. In this role, although you are not a CEO or divisional head who needs to put together the grand plan of the business, as a leader in your organization (and all data scientists should consider themselves thought leaders, even if you are not anyone's manager), you still need to understand the top-level strategic plan, and have some sympathy for how it is produced.

© Robert de Graaf 2019
R. de Graaf, *Managing Your Data Science Projects*,
https://doi.org/10.1007/978-1-4842-4907-9_1

However, if you are a data scientist in a consultancy, although you may skim some parts, you will still get a lot out of the overall theme of strategic alignment, so as to demonstrate the value of your work to your clients more effectively.

The Idea of Strategy

One of the key ways military strategy has been taught in the United States is according to the formula Strategy = ends + ways + means, with future military leaders using this summary to analyze past commanders' campaigns.[1]

Using this formulation provides an ability to analyze how strategies are varied according to the objectives and the means available to achieve those objectives, although I will often blur the distinction between ways and means.

Arthur Lykke offers some examples of military objectives in his article discussing ways and means. They include defend a homeland, restore territory, and deter aggression. It is intuitive that, for example, deter aggression could be achieved by projecting an appearance of powerful forces without firing a live round, but restore territory will require incursion into an area currently held by the enemy.

In his original article, the metaphor that Lykke paints is one of a stool with three legs, where each leg needs to be same length and angle for the stool to remain balanced (and therefore resist being knocked over). While we will more often use a two-sided metaphor, the point remains that the elements need to be balanced to ensure success.

Data analysis is a field where pursuing details is especially likely to be attractive, from which it follows that not seeing the wood for the trees is especially risky for a data scientist. Taking care to explicitly consider the big picture is one possible antidote to this problem and is the first step to knowing how far to chase down the detailed aspects of a data set.

Although data science has been promoted heavily as an answer to many of the world's problems, there are signs that at least some people have begun to feel that the promise hasn't been lived up to. A very plausible reason for this gap is that the overall objectives for data scientists, both as teams and within projects, may not be clear enough. This can cause a situation where data scientists are delivering projects that are excellent on the metrics the data scientists themselves set, but which fail to deliver the organization's goals.

---

[1] Arthur Lykke, "Towards an Understanding of Military Strategy," in *U.S. Army War College Guide to Strategy*, edited by Joseph R. Cerami and James F. Holcomb, Jr. (Carlisle, PA: Strategic Studies Institute, U.S. Army War College, 2001), www.au.af.mil/au/awc/awc-gate/army-usawc/strategy/.

This brings to mind another military adage, "War is too important to be left to generals," which is loosely meant as a warning that generals can win wars and battles without gaining anything of value for their country if they don't communicate well enough with their political leaders.

Winning Battles Without Winning the War

Military strategies have elements in common with data science strategies. It has been recognized that military means are often used in support of political objectives. This was starkly shown in an exchange during a meeting around the end of the Vietnam War between US Colonel Summers and North Vietnamese Colonel Tu when Colonel Summers observed "You know, you never defeated us on the battlefield," and Colonel Tu retorted "That may be so. But it is also irrelevant."[2]

The North Vietnamese strategy, which recognized that on the one hand the United States had unbeatable military strength while at the same time US troops could only remain engaged in Vietnam as long as there was political support in the United States, was effectively to avoid confrontation and wait the United States out. This has since been recognized as crucial to the success of the North Vietnamese.

This may appear to have little to do with data science. However, matching resources and capability to goals is critical to achieving those goals in any sphere. In the context of data science this may dictate what software you need, what training you need, and hiring decisions. For example, there's no point in doing a course in text analytics if you're not going to be able to use the new knowledge.

Alternatively, if, as was the case for the North Vietnamese, your capabilities are to some degree fixed, it means considering what projects you can take on carefully—at some level, even if you are an in-house data science department, success in the current project is needed to keep securing future projects, so accepting projects with a low probability of success has a strong chance of backfiring.

Tools: An Embarrassment of Riches

Data scientists are blessed with an abundance of different tools and a lot of time mastering their methods, but there is a lot less consideration put into the objectives. Intuitively, this risks disappointing customers and users, and there are now signs that this has been happening.

---

[2]Summer, Harry G., "Deliberate Distortions Still Obscure Understanding of the Vietnam War," Vietnam Magazine, August 1989, accessed via www.historynet.com/deliberate-distortions-still-obscure-understanding-of-the-vietnam-war.htm.

The flipside of such a large array of possibilities is that there isn't much agreement over what the standard toolset is. There isn't all that much agreement on what a data scientist is. Therefore, where there is always a tendency among experts to disagree on the best method to deal with a specific problem, this difficulty is especially acute for data scientists.

It's inevitably the case when data scientists can come from a statistics background, a computer science background, or some other background that doesn't instantly seem related. A strong sense of strategy offers a solution to this problem, as it allows you to define the concept of a data scientist that is most applicable to your team in your context—including both the skills available and the problems you are trying to solve.

This situation can be exacerbated by the need for statistical modelers to aim for very specific goals when they are modeling, evaluating the results by formal measures that do not always have a straightforward link between what the organization or client needs. It is incumbent on the data scientist to keep a firm grip on the customer's idea of value. This will ensure not only that the model or data product succeeds in providing that value but that the customer sees the value.

Frank Harrell, in *Regression Modeling Strategies*,[3] offers a number of strategies that effectively imply two orthogonal axes—one measuring predictive ability from none to superior, and the other measuring ability to explain the role of inputs from none to superior. He also suggests complementary strategies, where effectively separate models are created for prediction and inference.

This analysis provides low-level objectives (build a model that predicts well even if it can't be understood or build a model that can be understood, even if other models could be better predictors), but the higher-order objectives aren't in view. Note that Harrell's text offers fully developed strategies, in that after presenting objectives, he then outlines a method to achieve those objectives.

The different strategies from Frank Harrell are all modeling strategies. However, there are also other kinds of low-level strategies to consider. Models have different real-world uses. Sometimes they are used to make decisions. Other times they may be used to aid communication or to persuade an audience.

However, number of times the word visualization is used in articles for data scientists (referring to tools for producing graphs as a means of communication) compared to the word communication, or, even more so, words like convince

---

[3]Frank E. Harrell, Jr., *Regression Modeling Strategies, Second Edition* (New York, NY: Springer, 2015).

or persuade shows the focus on ways and means among data scientists over the ends they serve. Even more telling are articles where data science tools—the means to execute a data science strategy—are presented without a discussion of the objective they could be used to achieve.

The preceding strategy formulation also implies that there is a strong connection between available tools and objectives—strategy texts regardless of context often include examples of strategies where the objective is carefully chosen in recognition of minimal availability of resources—and often the ingenuity of a strategist is noted when they overcome a lack of certain vital resources in still achieving their objective.

Another area that is influenced by your overall strategy is which project management tools you will decide upon. While it is possible to choose a different methodology for each project, in reality, most teams stick with one methodology for general use and fit projects inside it. We will examine some of the alternatives in a later chapter, but the key takeaway is that an understanding of the overall team strategy is essential to picking the methodology that fits best.

The type of projects you select if you are an in-house data science team might also be influenced by the urgency of providing a pay-off. The notion of the quick win is very common and is often necessary to build credibility with upper management, but it will often come at the expense of bigger achievements, so if you know that you are being given more time, you can work toward something that will be more impressive at the end.

Understanding your goals and the resources you have to achieve them ensures that the goals are within your capability, a prerequisite for meeting them, and ensures that the goals you achieve actually assist your organization in achieving its goals.

A reasonable criticism of strategic planning as an activity is that too often the strategy is left on the shelf to gather dust rather than inform business decisions. However, the reality is that those decisions will still be made—individual data scientists and their direct managers choose not just which projects to pursue, but how much effort to invest in them.

Given the decision is unavoidable, you may as well base it on how useful the project is to company goals than more whimsical criteria, such as whether the project lets you use a cool new tidy verse package, or simply your emotional state on the day.

Getting this right means you can unlock a lot of additional value within your team and your organization, because as you are successful, and are seen to be successful, you get additional license and credibility that you can use to pursue more ambitious goals. Keep having successes, and you maintain that license and credibility within your organization.

Moreover, a clear strategy should provide clear links between goals and the actions taken to achieve them. This clear linkage is fundamental for anyone who wants to receive credit for their work, as it provides the means to explain how the low-level activities performed throughout any period contributed to the top-level goals.

Without those clear links, you may need a lot of goodwill and understanding from your audience to convince them that your team contributed to the extent that you believe.

The Value of Data Science in Your Organization

The idea that a data science function is beneficial to companies often seems to be assumed without proof. At least, the benefits are not well stated—it is assumed that whatever the data scientists do will instantly be of use to the company. In reality, there are a small number of types of ways at a high level that data science can be useful.

Stating what they are enables you to ensure that you align what your data science function is doing closely with what your company is doing. Just as it is easy to have a data science project that achieves all its objectives without helping anyone, it is also easy to have a data science function that produces fabulous work without improving any of the company's goals.

A hazard to developing a successful data science strategy is that a data science manager will be selected from among successful data scientists. As such, they stand a good chance of having an astute strategic mind, but may not have taken any formal study of corporate strategy allowing communication on an even plane with the MBA grads who are most often retained as consultants or in-house strategy gurus.

It was once said that war was too important to be left to generals—from the data scientist's point of view, it could be said that business management is too important to be left to the managers. Therefore, data scientists should educate themselves not just on how businesses work but on the vocabulary used to manage them, so they can communicate in the right language.

The most obvious company-wide use of a data science strategy is to research the company's customers, especially in a consumer setting where large numbers of customers allows for statistical analysis.

An inverted version of this is risk analytics, where rather than identify opportunities to market differently among statistically large groups of customers, similar statistically large groups of customers are analyzed to understand the possibility of their causing the business to lose money.

The personalities of people likely to be successful in these two areas intuitively could be quite different.

A third kind of data science is operational data science, essentially reducing costs. This is companies attempting to use data science to improve the way their business operates. This can be an approach to data science that lends itself to very creative approaches. It may include people looking deeply into logistics or people using deep learning identify the images in photos taken by drones, for example, photos of pole top transformers or overhead cables to automate the maintenance of these assets with reduced need for a human being to conduct the inspections.

There are also the goals of making decisions and choosing strategies. Here the data scientists are researchers advising senior managers on possible courses of action. The main tools in this arena will be optimization and decision theory, sometimes coupled with Bayesian statistics.

There isn't a mystery about these categories; however, there is always value in stating out loud what you might have already figured out subconsciously. As always, writing down what you think you know helps check that you really know it (amazing how what you think you know is right in your head can look or sound wrong when written down or spoken), and ensures that everyone else that you think knows the same thing really does know the same thing.

The process of getting the team together to create the strategy is as much about discovering what everyone was thinking all along as it is a process of having brand new ideas.

Strategic Alignment

Earlier I touched on the idea that a team's strategy needs to assist the wider organization in achieving its goals. In business school jargon this idea is called "strategic alignment" (knowing the jargon is useful when doing a Google search for more information).

Strategic alignment is recognized as a difficult area to master, although it is usually discussed from the point of view of senior managers trying to align their followers, rather than from the perspective of more junior people trying to align their own work upwards.

Another key concept in the way that organizational strategies are communicated throughout organizations is of "cascading strategies." This is where senior management creates a strategy that is reapplied at lower and lower levels of the organization, each time becoming more detailed.

In my experience, the underlying analogy is flawed in that a "cascade" still functions when the lower levels are passive. In my experience, for this process to function correctly, the lower management levels need to be ready to receive—like a pitcher and catcher.

Although there is usually some lip service paid to the idea that the senior manager at each level will discuss the strategy with their subordinate, too much emphasis is placed on the idea that the more senior manager will direct the conversation. In the case of a data science team strategy, where it will often be the case that the more senior manager is far from expert on data science, this is a dangerous path to take.

Overall, it is critically important that the data science function both serves the overall mission of the organization and is seen to serve the mission of the organization. To achieve this, a prerequisite is that the data science team's strategy is well aligned to the organization's strategy.

At the same time, you need to ensure that it is easy to show that the alignment exists, and an intuitive way to do this is to create a document showing this strategic alignment.

Why Alignment Matters

It's crucial for data science teams to be able to demonstrate their value to the organization by linking their activities to the organization's mission because there will be doubters who believe strongly that data science is a waste of time.

In fact, while it may sound paradoxical, it's still important when many people, including senior managers, are convinced that data science can make all the difference, as the kind of senior manager who has adopted data science because it's the latest fad will be especially fickle. Hence, for both the unconvinced and the completely convinced real results that are meaningful in their context are vital.

While the preceding points relate to the way your work is viewed within your organization—which can still often have real-world consequences—strategic alignment has been shown to be crucial to the actual success of an overall organizational strategy.

A well-known article in Harvard Business Review[4] by authors Bower and Gilbert showed that in many ways line management and functional management going about their everyday tasks did more to determine strategy than the activities of the most senior management. It may also be observed that many

---

[4]Joseph L. Bower and Clark Gilbert, "How Managers' Everyday Decisions Create—or Destroy—Your Company's Strategy," Harvard Business Review, February 2007, https://hbr.org/2007/02/how-managers-everyday-decisions-create-or-destroy-your-companys-strategy.

corporate strategies are vague enough to allow substantial room for an individual to put their own stamp on them.

It is therefore easy to implement the corporate strategy in your area in either a way that does uphold its intentions, or in a way that does not. Making a genuine effort to implement in a way that upholds the intended purpose is a clear way to gain credibility and influence with upper management, and make your functional area more important to the overall business.

Pro Tip   The examples used in the Bower and Gilbert article are also a powerful illustration of how personal influence can trump formal power. Their opening example talks about a factory that was built by line managers without senior management approval by purchasing components in increments under spending authority limits. In this scenario, someone had the idea and influence to get the others involved to go along with it. Maintaining and using influence as data science experts is crucial to ongoing data science success.

There is a very real sense that regardless of any data science strategy above you, the strategy you create for your team will be the data science strategy that is actually implemented in your organization.

Having this real power over your own direction gives you a great deal of autonomy, but it can be yanked away quickly if a perception that the company's goals are not being achieved ever develops. The way to prevent that happening is to ensure both that the work you do supports the company's goals properly and that senior people in the organization know that your work is supporting those goals.

Both of those things are far easier to achieve if the data science team has a clear, documented strategy that fully explains how the team's goals align with the organization's goals. The first step in developing a team strategy that achieves alignment with the organization is a careful study of the organization's strategy document.

Working With the Organization's Strategy Document

Most organizations renew their strategy at intervals and publish at least the top view version (sometimes the detailed roadmap may not be available to everyone) within the organization. It may be in a folder in a shared drive or on an internal company web page, for example. Get yourself a copy of this document, and analyze it.

Every organization has slightly different overall goals and mission. The strategy document describes what success for the whole organization looks like. That may mean chasing technical perfection, it may also mean doing social good, such as improving safety, it might be maintaining a high level of market share, or it may mean having the lowest costs. The challenge for the data science team is to decide how the data science team supports those goals.

Clearly, this will vary widely according to what the goals are, and what resources are available. In an organization driven by cost, the strategy might be centered on a search for automation opportunities. Alternatively, where the data science team is focused on risk analysis, the team's mission may be chasing incremental improvements on existing practices.

You may think that the connection between some or all of your team's usual activities and your organization's top-level strategy is both obvious, and well known to all members of your team. This is great—if it's true. It's often the case that incorrectly assuming people know something when they don't is more damaging than knowingly keeping them in the dark.

At the same time, if you stay with a business long enough, they will change their strategy, and if you don't you'll move to a new business—with a different strategy. In both scenarios, your ideas, if not the rest of the team's ideas, on what the mission is will need to change—just because the mission was understood on day one doesn't mean you can assume it remains well understood on any later day.

Having your team brainstorm ways to connect or answer the organization's strategy with you is still a useful thing to do, even if you feel that the connections between the top level and the team level are obvious. You may believe that each team member understands these connections, but only when you hear them say it do you know it.

You may also think that everyone agrees on what those connections are—but data scientists are experts, and any room of five experts has at least six opinions. Again, hearing people express what they "know" out loud means you know what they think rather than just guess.

Documenting Your Strategy

Strategies are complicated and neither easy to remember nor understand. It makes sense to create a record of what's in them to refer back to and to communicate their content to the intended audience.

Depending on the level of detail and complexity, the structure and content of these documents can be varied. A strategy document intended to be used for a team rather than an organization could be expected to take a relatively light approach. However, there are some parts of strategies created for larger groups of people that can still be useful at a team level.

Firstly, it can be useful to create a document that defines the mission and strategy for your team. These are common at the company level, but are sometimes forgotten at the departmental level—but if a department or team isn't very directly working on either selling the company's products and services or delivering them, a mission statement, possibly accompanied by team values can be useful to define how the team fits in the bigger picture.

The other crucial element of a strategy document is the plan itself, with details of the goals, how they are going to be achieved and who is responsible for getting there. Given a theme of this chapter has been the importance of aligning the team's goals with the organization as a whole, an intuitive addition is a statement on how each of the team's goals relates to the company's goals.

However, note that the same temptation to get lost in the details during analysis applies to a strategy for data analysis. A short document that is referred to often, with some items omitted, is more useful than a comprehensive tome that no one looks at because they can't ever find the section they're looking for. We will see that the simplest model for the job is usually the right one—it is the same for strategy and other similar documents.

The overall aim is to provide a short document that can be referred to often, and that leaves little room for differing interpretation. It should also be applicable to a range of possible data science projects, and help people understand both when and whether to do projects.

A Mission Statement for Your Team

The idea of writing a team mission statement isn't new—it's certainly encountered in Agile practices. Mission statements come in a variety of shapes and sizes, and although they most frequently encountered at the organizational level, sometimes they are useful at the team level.

This may be especially true for data science teams in large organizations that aren't intrinsically data science or similar consultancies. In this case, the team's mission statement allows you to align the team's objectives with the organization's objectives.

A good mission statement can be read aloud any time you decide whether or not to do a project—if it doesn't fit the mission, don't do it (if you have that power). If you don't have the authority to say no, write a business case for not taking on the work, even if sending it isn't feasible.

At the same time, the purpose of a mission statement is to unite people behind a common purpose. Just as important as a common purpose is a common way of doing things and approaching problems that allows your team to put a distinct stamp on your work.

The document will need to include details of how your team's strategy aligns to the top-level strategy. This may not necessarily mean that it fits to every point, but it should mean that it fits to the overall mission.

A mission statement can have four major components[5]—the mission, the vision, values, and major goals. You may feel now, and eventually decide, that your small team doesn't need such a large document. However, I'd say not to be too hasty about a couple of these.

When aligning the two strategies, consider what you are currently doing, and how well it fits to the organization's mission. Do you need to change? If you change, do you have the resources to make that change—will you need new data sources, or will you need to have new training?

A mission statement for your team, especially when created by the whole team together, can be an excellent platform for defining the way the team expects to aid the organization as a whole. Aim for something simply expressed, and make it as accessible as possible so that it gets used.

Presenting Your Team's New Strategy

A strategy document is no use if you keep it private. By involving your team, you've increased the chances that they sign on to the team's strategy, but it's still not guaranteed. You need to go through it with them and ensure that they really understand it, and know their part in achieving it.

Having a meeting with your team members is the minimum of what can and should be done to ensure that the strategy is understood. Be prepared to discuss it at the beginning of each project, and during regular team catchups. Make sure it's easily available to your team members, and encourage them to refer to it often—the best way being by doing that yourself.

To prepare for the meeting consider why you are doing this—answering both the *what* and the *why* is crucial to the explanation. Give background on the company's current situation. This is a good time to give a very brief update on where the company and the department are at (but not a good time for a detailed view that distracts from your message). That way, everyone will be hearing the new strategy from a position of equal awareness of the external situation. It also means that anyone who might have felt left out of things is brought on board at the same time.

While this recap on your organization's landscape needs to be brief, and therefore you need to be selective and tailor the talk to what matters at this

---

[5]Charles W. L. Hill and Gareth R. Jones, *Essentials of Strategic Management, Third Edition* (Mason, OH: South-Western, 2012).

particular point in time, there are some general issues you might like to consider.

You should expect questions, and you should be able to anticipate some of them. That is, it should be part of your preparation to brainstorm some of the questions you would ask, and to attempt to put yourself into the brains of a couple of your employees, and try to predict the questions they may ask in order to prepare some answers. Some of the things you may wish to put into the main presentation, others you may wish to leave, and only ask if they do come up as a question.

Pro Tip Assign different parts of your team to different team members who can be the champion for that section. For example, someone in your team could be the methodology champion or the data sources champion. This will get the buy-in of those particular team members, and mean that you don't have to go it alone in rolling out the strategy. However, you may not think it's a good idea to spring these roles on people at a positive meeting when you're trying to attract buy-in. Instead, have a one-on-one with those specific people, and use it as an opportunity to road test that element of the plan and the presentation.

Keep a close watch on your tone and overall presentation when you present this strategy for the first time. The point is to get your employees excited, if not as pumped as if they were at a rock concert. Rehearse your talk, and listen carefully to the tone of your voice. It needs to communicate excitement.

Consider the time of day where you have your most energy, and the time of day when you have the least. Don't schedule the talk on a day when you are running from meeting to meeting or when there is a distracting deadline to meet, or at least do as much as you humanly can to avoid those situations so that you have the time to do the presentation properly, and the time to take questions unhurriedly afterward.

Just as your team has a strategy, and, as we will see in Chapter 2, so does each project you attempt, so does your talk. Be clear in your mind what you are trying to achieve with this presentation. In this instance, merely informing is not the whole story—you are trying to get your team to sing on with you for the way you will be doing things for the next 12 months (or your preferred time frame).

Overall, how well your presentation is received at its first public airing will rely on how well you prepare, and how well you can empathize with your own staff to ensure you have the right answers at your fingertips.

Putting a Strategy into Action

A strategy has very little—if any—value until it has been used to make a decision. The size of the decision isn't important, but at some point it should be expected that the strategy is used to guide a decision.

Typical decisions in a data science life cycle could include choice of algorithm; choice of data; how much time spent on data cleaning; how much time to spend on feature engineering. While questions like which algorithm to use are part strategic, part technical, questions that begin with "how much time/ effort to spend on X?" can be seen as almost totally strategic.

Another important time to go back to the strategy is during the hiring process. I've met recruitment agents whose workaround to the problem that neither companies nor candidates truly know what a data scientist is has been simply to ask "How do you define a data scientist?" to anyone they speak to. A clear strategy that supports a strong understanding of the skills you need on your team side steps that problem.

It's important to get this right, because on the one hand senior management is unlikely to include people with a deep understanding of data science, at least not for some years from now. Hence, it will be up to the managers and team leaders to translate the overall company strategy into a strategy for the data science team.

On the other hand, a frequent reaction from senior managers where there appear to be problems is to manage more. That is, if results aren't achieved rapidly, or at least not rapidly enough, a plausible outcome is that senior managers will try to involve themselves more and give more instructions, regardless of whether those instructions make sense in a data science context.

Pro Tip The comment that interference can lead to worsened results isn't intended as a dig against senior management, but rather as a comment on human nature. Most people, if they see something going wrong want to fix it. That they are at risk of worsening the problem comes from the frequent warning[6] found in Statistical Quality Control manuals that unnecessary adjustments frequently lead to exactly the poor process outcomes they are intended to avoid.

If the instructions are inappropriate, this can lead to a vicious circle where the team performs progressively less well, while the senior managers try to manage more, making the problem worse.

---

[6]Berger, R.W., Benbow, D.W., Elshennawy, A.K., Walker, F.W., *The Certified Quality Engineer Handbook* (Milwaukee: SQ Quality Press, 2002).

All of this can be avoided by setting goals for your team that relate to the organization's goals, but use words that make sense to your data scientists. Partly because humans have a tendency to forget and to break promises, and partly because data science is usually a group activity, it's important to write these words down.

When making this strategy, be aware of the need to build support. We will see in future chapters that achieving the best results is often dependent on getting good advice and information from the people in the areas you are trying to assist, as well as from other subject matter experts. Building a pipeline of projects that provide early success can be instrumental in ensuring you continue to enjoy goodwill from this group of people, who are vital to your overall success.

Showing You Mean It—Behaving As If Your Strategy Matters

Nothing kills a leader's message as quickly as the leader's followers seeing the leader acting against their own message. Conversely, the most powerful way to communicate a strategy is to act on it in clear view of your followers.

When you are assigning tasks to yourself, therefore, make a point of telling the team what you are working on as much as possible and also make a point of explaining how your work fits into the overall strategy.

Also make sure that you show that you are tying your decisions back to the strategy across your usual activities. Mention that it's the reason behind picking a particular project or avoiding another, and make it a part of performance reviews.

Giving your team the idea that you don't care about something is a shortcut to them not caring about it either. It's not a secret that actions are louder and easier to hear than words. It's not different in relation to a team strategy than it is anywhere else—if you want people to do something, the best way is for you to start doing it yourself, somewhere you can easily be seen.

Strategy and Culture

Throughout this chapter, we have concentrated heavily on written strategy. However, nonwritten strategy, usually conveyed as a shared culture, can be a critical way to maintain and communicate strategy. For a strategy to succeed as a decision-making tool, for example, reaching a point where the culture within the team makes decisions aligned with the strategy nearly automaticaly creates the best results.

However, the details of how this is done are difficult to capture in a document. Possibly the most famous attempt to document culture in a working document was performed by Jews under the Roman occupation. Recognizing that their written record of religious and cultural practices, the Torah, covered only a fraction of those practices, they attempted to write down their oral law, and created the Mishnah, ostensibly the "Oral Torah," which was far longer than the "Written Torah" (the first five books of the Jewish Bible or Christian Old Testament).[7]

While clearly documenting the culture of an entire society is a far greater task than for a functional team within an organization, the point is that creating a document to define culture may be too large a task.

Instead, consider the use of routines and rituals that create the right culture for your team. These will need to suit your team's makeup and context, and are better invented in conjunction with your team to make them "sticky."

Also consider how to incorporate some of the other teams that you work most closely with into the data science team's culture. As one of the groups in an organization with some of the most specialized training, it is especially easy for data science teams to become isolated from other areas, leading to a "silo" effect.

The most obvious remedy is to interact more with the teams around you, and it may sometimes be easier to do that within organized or semi-organized social occasions than during your day-to-day work. Talk to the people around as well as the people inside your team to get the maximum effect.

The benefits of incorporating the other teams around you are especially important, considering the need to understand the organization's goals as deeply as possible. It's inevitable that people in those other areas will have a different idea of what those strategic goals mean—indeed, this is intuitively a reason why executing on a top-level strategy is so often very difficult.

Maintaining links with the other teams is crucial to ensuring that you are able to keep tabs on what the top-level strategy means to other people. It, therefore, ensures you avoid aligning your team's strategy to an interpretation of the top-level strategy that others in the organization won't recognize.

---

[7]James Fieser and John Powers, *Scriptures of the World's Religions* (New York: McGraw-Hill, 2008), pp. 267–275.

Prepare for Friction—Overcoming Obstacles to Your Success

A great deal can go wrong when delivering a strategy. Some of the obvious things that have been covered include poorly communicating the strategy, and poorly aligning the strategy with the organization's goals. Obviously, these are just a couple of obstacles you may encounter.

One of the assumptions of this book is that there are organizations out there right now which have incorporated data science into their overall strategy but will not get the results they want due to a poor understanding of what data science can and cannot deliver.

One of the most difficult and most universal pieces to strategy implementation is change management. A strategy that requires no change is unlikely to achieve anything, and unfortunately, as all employees including data scientists are humans, there will be people who do not want to change.

There are numerous models of change management, and you may think that some of them are over-specified for the job of getting team members to buy into a strategy, that is, their job to buy into. However, using a small number of common sense measures can still make a big difference:

1.   Ensure you explain the reasons for the new strategy.

2.   Address any concerns that arise as you explain.

3.   Maintain a positive attitude (you are senior management's ambassador).

4.   Have some of the answers to hand on things that will be needed to achieve the strategy, such as training, new tools, or new data.

Although ultimately people need to sign on to a new strategy because it's their job, there are still many things that can be done to make the process easier. Like many things in life, a lot of these come down to knowing what to expect and preparing accordingly.

Change management as a process is one of the most studied aspects of management and leadership, and as a result, there are many resources available. As a data scientist, mastering this particular skill will be crucial for your ability to lead not just your own team but often to how well your models are received and used in the wider world.

We will return to this point multiple times throughout this book—how much your models are used and valued by your customers will very frequently depend less on how effective the model itself is, but on how well you manage the adoption process. Hence, although this skill has been introduced in the context of how to properly introduce your team's strategy, in reality, your skill in this arena has much wider implications for your overall success.

On a final note, research[8] has shown that one of the most significant reasons that strategies fail is that they are developed by senior management who are then unable to directly verify the implementation—instead they are reliant on their middle managers to communicate the results.

In addition to the simple actions listed previously, there is no doubt that one of the most effective ways to assist with the process is to be seen to embrace the change yourself. Compared to a senior manager without the advantage of working alongside many of the workers in an organization, you have an advantage that you can be seen actively implementing the strategy on a daily basis.

If you are a data science team leader or a data science functional leader, you are in a more fortunate position of being able to see directly how the strategy is implemented on the ground. Although you should take the challenge of change management seriously, there is a lot of reason to feel optimistic about implementation.

Close the Loop—Check Your Results at the End

A common mistake is to produce a strategy, and then fail to evaluate its success—or otherwise. In contrast, by returning to the goals you set out to achieve at the beginning and evaluating your performance, you ensure that the lessons of what went wrong and what went right aren't forgotten. At the same time, you can adjust your view of what you are capable of, and what your goals should be.

Somewhere along the way, you made some mistakes, so you should reap their value. Somewhere along the way you realized that there was a better target than what you were actually aiming for—again you should adjust. It was expensive in both cases, so don't let yourself be short changed by not getting the full value of those mistakes.

Meanwhile, circumstances have changed since you wrote the original strategy. Different tools are available, and probably different data are available. If not a complete new source, then at least you now have a greater sample of the kind of data you were using originally, as data accumulated while time marched on.

Your team, too, has changed—they are more experienced, and hopefully more skilled and a little wiser. It should be possible to demand more of them than it was at the beginning of the process.

---

[8]Syrret, M., *Economist Guide to Strategy Implementation* (Profile Books, 2007).

On the other hand, maybe some team members moved on, and the overall team composition changed. Take the opportunity to reassess your strategy in light of the overall team's new skill set.

Keep a record of these sessions, and they will become valuable resources, not just as an input to the next round of strategy creation, but as a generalized "lessons learned" library. You wouldn't throw out your code snippets, or data sets, you'd archive them. Do the same with these lessons learned.

A Strategic Thinking, Planning, and Doing Life Cycle

A strategy, like a model, has a useful life beyond which it is used up or wears out. The final step in the process, the review step, helps you decide when this has happened and sets you up to get the maximum benefit from what you learned applying the old strategy when you sit down to begin to write the new strategy.

Aside from the lessons learned that relate to how well the strategy was executed, there is also the question of how fit for purpose the strategy itself continues to be, partly determined by any changes that the organization's senior management have made to the top-level strategy.

There may be an automatic process in your organization that reviews that top-level strategy and, therefore, prompts you to look at the team strategy. There may not.

Where there is a process, you can get the most out of it by using it as a prompt to realign your team strategy to the top level. If there isn't a process, make your own process, and put aside time on a regular basis to check whether your strategy still supports the top strategy.

That time is also an excellent opportunity to share some of your wins. Although your organization most likely has a formal performance review process where you are expected to account for yourself over the year, they are a poor forum to let senior management know your wins. Moreover, as they are one-on-one, you lose the morale benefits that come with praising your team members in front of more senior managers.

Take the opportunity when you do your own review to compile highlights of what you've achieved, and present it to senior management. This allows you to achieve the final important benefit to having a well-defined team strategy in alignment with the high-level strategy—the chance to link your work with the organization's success in the minds of the organization's top managers.

The chance to promote your team's work and highlight its value is what this has been leading up to. The process of creating a well-aligned strategy, allocating time to the projects that support it the most, and reviewing your progress against your initial goals ensures that when you do so, the result can only be that your team is respected and credited for the value they bring to the organization as a whole.

The review process, which completes the cycle by checking the outcomes against the goals and feeding the results back into the planning process for the future iterations of strategy planning, is the essential last step that ensures that the value of your team's efforts is fully realized.

Summary

According to Herb Kelleher, founder of Southwest Airlines, "Strategy is overrated. We all have a strategic plan, it's called doing things."[9] That may be true, but it's unlikely that by fastening pieces of wood together with no guiding principle that a house will naturally emerge. As noted earlier and reaffirmed by the preceding quotation, the decisions that make up a strategy are made on a weekly, daily, or sometimes hourly basis. Without a strategy the reasoning will be local, and will often be arbitrary and isolated from the other decisions. With a strategy, those decisions can be validated against a higher purpose.

A data scientist embedded in an organization that is not primarily a data science consultancy needs to align themselves with their company's overall strategy—there is a task of translating the goals of the company as a whole into goals within the data science function which support those goals. That process is crucial to the success of the strategy.

Even data scientists who work in consultancies can benefit from a greater understanding of their clients' larger goals. Every project has a context—even if the client understands you did an excellent job on a project, it can still feel tainted on their side if it was a poor fit for their overall goal.

The resultant data science functional strategy is both a decision-making tool that informs people on both what work to take on and how much importance to place on it. At the same time, it is a communication tool that ensures you can explain how each project and the department as a whole are supporting the overall business.

---

[9]Although this quote is frequently attributed to Herb Kelleher and he most likely said it on various occasions, the only recorded source I could find is "When planning became big in the airline community, one of the analysts came up to me and said, 'Herb, I understand you don't have a plan.' I said that we have the most unusual plan in the industry: Doing things. That's our plan. What we do by way of strategic planning is we define ourselves and then we redefine ourselves." www.strategy-business.com/article/04212?gko=8cb4f.

Any strategy's success depends not just on its quality as a strategy but how well it is communicated. Involve your team with its implementation and keep the way that it is expressed as simple as possible.

As senior management are seldom data scientists themselves, they aren't able to stay on top of what's happening in your area at a detailed level. It is therefore not to be expected that they are able to provide a strategy that is perfectly aligned to your function.

For best results, you need to have a certain amount of self-reliance, but self-reliance that is sympathy with the organization's goals, rather than in pursuit of your own goals. In this way you have a strong foundation to fill the gaps in data science strategies that more senior generalist managers cannot.

Of course, more than that, for a strategy to be effective it has to be used. Make it a habit to think through decisions in terms of your strategy. When deciding which of two projects should go ahead, or which should get the most time or deserves the most experienced person on your team in case her dance card is full, consider which project advances your strategy most effectively.

We will see later in this book that models need to be maintained to ensure they perform as expected throughout their life. The same is true for strategies. Not only that, reviewing a strategy provides a great opportunity to learn from your experience, and to make sure that what you learned is captured.

There are people who argue, similar to Herb Kelleher, that it isn't necessary to have a formally documented strategy. While the necessity may be arguable, having a simple team strategy that aligns well to company goals is a great way to ensure that your work not only supports those goals but also can be seen to support those goals.

It may be that there are occupations where it's instantly clear how they support the goals of the organization as a whole. It's intuitive that airline mechanics and pilots may know easily how they fit within the structure of an airline and support the organization's goals (although this doesn't mean it's guaranteed).

Data science isn't an occupation where it's instantly obvious how the contribution is made—let's face it; there are nearly as many definitions of data science as there are data scientists. Expressing your particular definition out loud and writing it down will pay dividends in added clarity.

A strategy has a number of parts. They can include a mission statement, a strategic plan covering details of intended goals and how they are to be met, or a set of cultural practices that have the effect of leading people to answering certain problems in a certain way. Understanding and untangling these elements gives you a platform that will allow you to not just achieve more but also to receive your fair share of the credit for achievements, and therefore continue to be given opportunities to do innovative work for your company.

No one has enough time to do everything that appears worthwhile, both as individuals and from a team perspective. Understanding the company's goals yourself and ensuring everyone in your team understands them, combined with a knowledge of how your data science activities support those goals, ensure that every ounce of your effort improves your company, and that you can explain to anyone outside your team how you contribute.

In Chapter 2, we will move closer to the metal by discussing strategy for individual projects. Although some of the philosophy will carry over, we will still see a great number of new ideas.

Team Strategy Checklists

In order to cover all the important aspects when developing a strategy, it can pay to have a short checklist to consult. The following is a nonexhaustive list of useful questions to ask yourself about your team's strategy to ensure everything important is covered.

Team Context

- How much goodwill currently exists? The level of goodwill often determines the timeframe within which you are supposed to deliver

- What skills currently exist in the team? Are they a good match for the data that is usually available?

- Where does data mostly come from? Do users or clients bring it to you, does it currently exist within the organization, are you expected to effectively obtain data for the organization, for example, via web scraping?

- Following on from the above, will you usually be looking at the same data set or looking at different data sets depending on the occasion?

- Is there an accepted industry standard way of looking at problems, for example, is a generalized linear model considered the "gold standard"? Is there an advantage of a new approach?

- How accepting of new approaches is your industry generally? What sort of blowback will there be if you introduce something new?

- Are there resources available in your organization to help educate users, or will that be your responsibility? Who are users likely to ask for help if they encounter a problem?

Alignment

- Can I use the team's strategy to explain how data science helps the organization achieve its goals?
- If I follow the team's strategy, will I automatically achieve the organization's goals?
- Will the organization's strategy be stable over the intended period of the team's strategy?

Strategy Documentation

- Is your strategy document easily available to your team?
- Does the formatting of the document make it easy to read?
- Have you made a point of communicating the new strategy with your team members and discussing how it relates to their individual work?

Presenting Your Strategy

- Choose a time when both you and the team will have as few distractions as possible
- Anticipate the most likely questions and prepare responses
- Speak one-on-one with champions for particular aspects of the strategy prior to presenting it

Culture

- What are your team's rituals? Do they help or hinder in achieving the team's goals? Do they make the team open to change, or do they reinforce a team filter bubble?
- Do your rituals include the teams and others around you?

Acting on Your Strategy

- Is it standard practice to consider whether a new project under consideration supports the team's strategy?
- Do you adjust the descriptions in job ads to attract candidates whose skills support the overall team goals?

Data Science Strategy for Projects
Meeting the Right Targets

In Chapter 1, we discussed strategy for data science teams. Yet you can't benefit from a strategy until you have achieved something for your customers. For a data scientist, this usually means completing a project. The strategy for the team guides some of the projects and its prioritization and is the base of the project's strategy. However, projects still need their own strategy to properly define their objective and the means that are available and allowable to achieve the objective. By nailing this at the beginning of each project, you ensure that each one achieves its objectives, and, crucially, your work receives full recognition from your customers.

Project Management

It is often considered that good project management is vital to having a successful project result. You can't simply start a project without a method to work out how to organize the work and how to estimate the likely time and

© Robert de Graaf 2019
R. de Graaf, *Managing Your Data Science Projects*,
https://doi.org/10.1007/978-1-4842-4907-9_2

resources required to complete your project and achieve your hoped-for result.

Over the years, many different processes have been used to manage projects. These processes all have in common that they systemize an approach to deciding what to do, estimating how long it will take to do it, and then ensuring that what is done is what the customer wants. However, they differ in the means of achieving those goals.

In the recent past, many software companies and software-oriented companies (i.e., companies where the main product is not software, but software is critical to the product's delivery) have adopted Agile as a key project management methodology. Agile is usually in contrast to a formal "waterfall" approach, where requirements are defined early in the project, and then changes to requirements are difficult.

The waterfall approach was pioneered for complex civil engineering and construction projects where the final desired outcomes are heavily dependent on earlier stages and change is very expensive. For example, consider how the floor plan of a building determines the foundations needed; once the foundations are poured, going back and making a change would be extremely difficult and expensive, and so, a change to the floor plan only be made that doesn't require different foundations.

As Agile has taken off, sometimes it appears that using Agile principles have made the frameworks used in waterfall projects redundant. However, there is a risk of throwing the baby out with the bathwater in carelessly or naively implementing any specific project management framework without considering the messages and strengths of the others.

Two myths busted by lean software development[1] pioneers, Mary and Thomas Poppendieck, combine to illustrate the point. The first is "Early specification reduces waste" and the second is "Planning is commitment."

These myths are ideas from traditional project management which have often been used to promote specifying early and creating a plan that is committed to. These are important things to do if you need to begin digging a foundation 6 months ahead of delivery of a piece of equipment.

However, the tools to make the plans and decide the specifications can still be used while maintaining the attitude that the plan can change. An early specification doesn't have to be handcuffs, it can simply be the results of your research.

---

[1] Mary Poppendieck and Tom Poppendieck, *Implementing Lean Software Development* (Upper Saddle River, NJ: Addison Wesley, 2007).

Therefore, by accepting that planning is not commitment, we can use the best tools from traditional project management methods, with as much of an Agile mindset as suits our situation.

In the material that follows, where we borrow material from waterfall project management, such as recommending creating documents and recommending talking to customers early, we are not demanding a commitment to the early discovery, simply observing that chances engage with customers may not come smoothly throughout a project life cycle, so use them where they arise.

One of the important features of many of the traditional project management exercises is an emphasis on formal opening and closing procedures in order to ensure that the right path is followed—an approach that emphasizes risk and quality management.

In traditional approaches, such as Project Management Body of Knowledge (PMBOK)[2] and Projects in Controlled Environments (Prince2),[3] this work takes place up front and produces a set of documents that record the results of the discovery. The disadvantage of this approach is that a rigid Project Manager can use these documents as a straitjacket, refusing to allow them to be altered in the face of changing circumstances or new information.

The advantage is that objectives are clear from the beginning. At the same time, there is no need to stop being Agile—simply take the best of the other approaches, learn from it, and apply it to your own situation. Indeed, the idea of hybrid approaches is one gaining greater currency today. Indeed, both PMBOK and Prince2 have developed guides on how to use those approaches while being Agile.[4]

As we will see next, choosing the right objective and defining it correctly is a difficult task. It can also be expensive when it goes badly. In that context, it makes sense to take learnings from as many areas as possible with a view to avoiding that expense.

Ultimately, there is nothing about trying to understand the customer as deeply as possible as soon as possible that conflicts with Agile principles, as long as it is done right. Rigidity in the face of changing requirements stems more from human defects than it does from defective processes, and isn't a reason to abandon a process.

---

[2]"PMBOK Guide and Standards," Project Management Institute, accessed April 8, 2019, from www.pmi.org/pmbok-guide-standards.
[3]"What is PRINCE2?" Axelos Global Best Practice, accessed April 8, 2019, from www.axelos.com/best-practice-solutions/prince2/what-is-prince2.
[4]See, for example, www.pmi.org/pmbok-guide-standards or www.axelos.com/news/blogs/october-2018/using-agile-project-management-a-hybrid-approach.

Defining the Objective

In many ways the most difficult aspect of data science in general, and of any project, is choosing the correct objective. In the case of data science, as a mathematical discipline or at least a discipline inhabited by people whose background in areas such as computer science and statistics makes them prefer to see problems in a quantitative light, practitioners need objectives with a numerical definition.

In contrast, it can be argued that in the context of framing a relevant business problem, a qualitative viewpoint can often be more helpful in the initial phases. For a start, it is important to understand what kind of answer the customer is looking for. Are they going to use your work as decision support, for example, to choose between some alternative courses of action?

Are they instead going to use the result in a quantitative way themselves, for example, to determine the number of resources to allocate or stock level to hold?

In many cases the output of the model is not the final action—it only contributes to the final action in some way. The actual final action is an implementation of the model or a decision based on the model's output. For example, although the output of a regression model is a set of coefficients to define an equation, and the output of applying that equation to new inputs is a column of numeric values, the true output is the decisions that are made as a result of understanding those values.

Taking a more concrete example, correctly estimating the person-hours taken to do something may be an obvious choice for a model, but it may not be the ultimate goal. The goal may really be to estimate the elapsed time (when can I have it?) or labor resource (how many people should I allocate?). Hence, choosing the right dependent variable to model may not be obvious, and there may be another variable, easier to model that is also more apposite to the client's needs.

Another side to this coin is understanding why the customer wants a data science solution. Do they want higher accuracy forecasts than they currently achieve? Or would they be happy with the same accuracy but want to leverage computational speed and power to achieve a faster turnaround time?

This last question should be key to an effective data scientist's approach, as clearly an hour or two spent ensuring you have the correct understanding of the client's objective will be more useful than 10 hours spent optimizing a model serving a different objective.

Costs of Getting It Wrong

Partly due to data science being a relatively new activity, compared to other disciplines, few examples of the cost of getting data science wrong have been compiled. However, there have been costs of failure analyzed in the overlapping context of software engineering. Some of these are glossed by *Code Complete 2*,[5] itself a brilliant resource for data workers resolved to avoid wasted time, despite the book's focus on software construction.

McConnell, the author of *Code Complete 2*, combines the results from a selection of papers to give potential ranges for the time taken to fix defects at different stages of software projects—not altogether surprisingly, fixing an error in requirements after the requirements stage quickly blows out as projects progress, until fixing a requirements error post-release is estimated at 10–100 times as time-consuming as fixing it during the requirements process.

Coming more from within the statistics world, Howard Raiffa[6] has called solving the wrong problem precisely a Type III Error (following the Type I and Type II errors familiar from introductory statistics classes). Mitroff and Silvers have collected a number of examples of this error (and divided these errors into intentional and unintentional, although we are primarily interested in the unintentional). While many of their example problems are too complex for retelling here, the following gives a flavor.

> The Manager of a large high-rise office building was receiving mounting complaints about the poor elevator service. She decided to call in a consultant to advise her on what to do to solve the problem ... [the consultants suggest expensive engineering solutions] ... fortunately, one of the hotel tenants was a psychologist.[7]

The psychologist succeeds where the elevator engineers fail by realizing that the key reason that people complain about the elevator is that they are bored. She installs mirrors near the lifts to keep the lift users amused, and the problem vanishes (and the authors note that had they invented the problem a few years later, it might have been TV screens, not mirrors).

The moral of the story is that solving the right problem is crucial. The problem was never that the lift was going too slowly, but that the lift's users didn't have anything to do while they waited for the lift.

---

[5]Steve McConnell, *Code Complete 2* (Redmond, WA: Microsoft Press, 2004).
[6]Howard Raiffa, *Decision Analysis: Introductory Lectures on Choices Under Uncertainty* (Reading, MA: Addison Wesley, 1968).
[7]Ian Mitroff and Abraham Silvers, *Dirty Rotten Strategies: How We Trick Ourselves into Solving the Wrong Problems Precisely* (Stanford, CA: Stanford University Press, 2010), p. 34.

The error that the engineers made of trying to speed up the lift rather than relieve the lift users' boredom is an example of a Type III error—solving the wrong problem with a precise solution. In data science this could easily happen if you create a highly accurate model with the wrong target.

More insidiously, there may be multiple "right" targets and multiple ways of building accurate models for them. The multiple right targets may differ in terms of the data they require and the tools they require to model that data. Therefore, there is a great benefit to the data scientist who can identify the right target that offers the most tractable solution. Sometimes, you won't need a better data preparation tool but a different way of looking at the problem which reframes it with a less onerous data challenge.

The multiple ways of building accurate models can also determine a lot about what sort of solution is offered to the client. For example, many people have observed that the super accurate models made to win Kaggle competitions are very different to the leaner models commonly found in industry, especially as the processes used to achieve the last 0.1% of accuracy that wins the competition are usually too computationally intensive to get an answer to the customer in a reasonable time frame.

Most data science professionals realize that the massive stack of advanced neural nets and Gradient Boosting Machines that win Kaggle competitions is unsuitable for most real-world clients' needs.

The more subtle area is where there are multiple models that could be feasibly implemented, and poor understanding of the customer's preference means that an optimum model could be selected that actually isn't optimum in the eyes of the customer. This is an example of Project Risk, in that it is a risk that threatens the perceived success of the project, but it is not one of the most commonly discussed project risks.

Project Risk and Unintended Consequences

Project risk is often defined and examined from the point of view of risks to the completion of the project. That is, in my experience, a typical project risk discussion will focus on hazards to completing the project either on time or correctly. For example, we will discuss the CRISP-DM data mining approach later in this chapter—it describes risk in terms of things "that might delay the project or cause it to fail."[8]

This definition is natural when considering projects that begin with a very strong definition, as, for example, civil engineering projects where the project

---

[8]Peter Chapman, Julian Clinton, Randy Kerber, Thomas Khabaza, Thomas Reinartz, Colin Shearer, and Rudiger Wirth, *CRISP-DM 1,0: Step By Step Data Mining Guide*, (SPSS, 2000), www.the-modeling-agency.com/crisp-dm.pdf.

team is constituted to build to a structure to a prescribed blueprint might be. However, in data science, and in fact in many software engineering contexts, the objective is far murkier.

A more insidious form of project risk is similar to a Type III error—the project is completed on time and correctly but doesn't give the customer the expected benefits. Worse than that are cases where the project is completely successful, and then inadvertently causes problems. This could be a Type IV error, although Mitroff and Silvers have already proposed a definition of Type IV error.

The possibility of adverse outcomes arising because of unintended consequences is an aspect of the safety of machine learning systems.[9] This is a new consideration in machine learning, but likely to be of increased interest in the near future.

A recent example illustrates the hazard of unintended consequences. An Australian woman wrote an open letter to tech companies involved in targeted advertising after her social media became inundated with baby-related ads after she bore a stillborn child.[10] In her open letter she posed the question "If [targeted marketers] could identify she was pregnant, couldn't they identify that she had had a miscarriage?"

Intuitively, the marketers are able to identify women who have recently had a miscarriage (and they could probably even market, for example, counseling services to them) but had not identified the need. While it's only possible to speculate on what the marketers did or didn't do, not deliberately considering when a wrongly placed ad can have negative consequence is a plausible reason why they didn't identify this potential problem.

Big Data—Big Risks?

The trend toward Big Data offers another area where attention to the problem being solved is crucial. Although it is true that Big Data can offer solutions that are not available with a smaller data set, there is also a widening realization that a larger data set also brings greater risks. In particular, data sets with many possible input variables bring an especially high risk of falsely identifying an input variable as important.

---

[9]Kush Varshney and Homa Alemzadeh, "On the Safety of Machine Learning Systems," *Big Data*, Volume 5, Number 3, 2017, https://arxiv.org/abs/1610.01256.

[10]Wendy Tuohy, "Didn't You See Me Googling Baby not Moving: Mother's Emotional Plea," *The Age*, December 12, 2018, www.theage.com.au/lifestyle/life-and-relationships/emotional-plea-to-spare-bereaved-mothers-from-hurtful-online-ads-goes-viral-20181212-p50ltv.html.

Therefore, if there is a way of answering the customer's needs (find more customers, better identification of risk, etc.) without using ever-larger data sets to achieve the result, this will often be a better outcome—and this is before considering the extra effort in computation and coding time that is usually associated with Big Data over more moderate data.

Finally, the biggest problem with Big Data is that it encourages people to focus on the fact of the data being big at the expense of a clear understanding of the client's problem. As we discussed and will reiterate further, a clear understanding of the client's problem should always be the central concern of any data scientist making a genuine attempt to provide value.

Defining the Objective

There are a number of approaches to elicit the best understanding of the customer's real requirements. We will review three of them, and discuss the context of how they are usually applied in their relevant contexts.

Each of these approaches has in common that they exist to widen the discussion from what people would automatically do if allowed to default to problem solving by reflex—human nature is to suggest a solution as soon as possible, without slowing down to discover the real problem, or at least, without checking to make sure that the real problem has been identified.

Hence, each of these approaches exists as much as anything else, to force people to slow down and hasten more slowly in their approach to problem solving.

The Six Sigma[11] process was originally developed for use in a manufacturing environment. However, after success, especially at GE, which began as a manufacturing company but branched into other areas including finance, it began to be used for a wider range of applications.

The Six Sigma approach is stage-gated, with the stages in the original version being remembered through the mnemonic "DMAIC"—Define, Measure, Analyze, Improve, and Control.

Fundamental to the success of the Six Sigma approach to solving problems is the lengths that are gone to understand the voice of the customer, which occurs during the initial "Define" stage. This is then translated into a measurable target that is the focus of the Six Sigma team. There are two important outcomes with this approach.

Firstly, it ensures that the subject of the project is genuinely relevant to the end user. Secondly, it ensures that people maintain their understanding of

---

[11]"What is Six Sigma?," accessed April 9, 2019, from www.isixsigma.com/new-to-six-sigma/what-six-sigma/.

whether the measurable target is the actual thing the customer is interested in or a proxy, and if the latter, ensures that the way the proxy and the customer's concern relate is transparent.

So how can we properly establish what our customer actually wants? The cold reality is that all too frequently they won't be able to tell us, although this may not mean that they don't actually know.

Although the Six Sigma was developed for a different environment than is seen in some data science projects, there is still a lot we can from this approach. This, especially, because arguably Six Sigma's greatest achievement was taking pre-existing quality assurance tools, and pairing them with a rigorous approach to understanding the real objectives of quality projects.

This link enabled Six Sigma users to ensure that they could explain their successes to the rest of their organizations—a vital consideration in most companies in the modern age, where if the management can't see you adding value, you can be very quickly removed from the business.

The first notable feature of the Six Sigma approach is that the Define stage—where the goals are set and success is defined—is given its due as the true foundation of any project. At the same time, there is no assumption made that the customer or client will be able to express their needs in terms that easily lend themselves to developing the clear targets required for this kind of project. Instead, various tools are used to convert what the customer knows they want or need into something more tangible lending itself to specific and achievable goal setting.

The tools that are typically used in a Six Sigma context are intended to help practitioners zero in on the part of the problem that has the most influence over the end result, or to put it another way, the area of the problem that has the best effort to benefit ratio.

In this book, it's not our aim to give a comprehensive or any guide at all to Six Sigma design tools. We'll look at just one to demonstrate the philosophy. We also note that, in general, tools and techniques claimed by Six Sigma were not invented for use with Six Sigma projects—they usually already existed and were identified as fitting the philosophy sometime after their use became widespread, even if being recommended as a Six Sigma tool made their use more widespread still.

The Voice of the Customer

Fundamental to the success of the Six Sigma approach to solving problems is that the objective is defined by voice of the customer. Only by understanding the voice of the customer can you find an objective to be translated into a measurable target that becomes the focus of the Six Sigma project. As a result, a substantial part of the Define stage is devoted to understanding the voice of the customer, and understanding the voice of the customer is recognised as the first step towards defining a project. There are two important outcomes with this approach.

Firstly, it ensures that the subject of the project is genuinely relevant to the end user. Secondly, it ensures that people maintain their understanding of whether the measurable target is the actual thing the customer is interested in or a proxy, and if the latter, ensures that the way the proxy and the customer's concern relate together is transparent.

So how can we properly establish what our customer actually wants? The cold reality is that all too frequently they won't be able to tell us, although this may not mean that they don't actually know.

In an ideal world, we might want to use a tool that Six Sigma practitioners use, or something very similar. However, we are at a disadvantage because Six Sigma practitioners are able to train their (usually internal) customers to expect certain tools, and unfortunately the expectation has grown that data scientists will march in, make and implement some models, and march out leaving easy profits in their wake. That doesn't mean we can't learn from the thinking behind some of the Six Sigma models, however.

The tools that are typically used in a Six Sigma context are intended to help practitioners zero in on the part of the problem that has the most influence over the end result, or to put it another way, the area of the problem where the ratio of the ease of fixing compared to benefit of fixing provides the most favorable results.

In this book, it's not our aim to give a comprehensive or even any guide to Six Sigma design tools. We'll look at just one to demonstrate the philosophy, and how it can work in practice. As always, and as will be the case for other tools presented in this chapter, the correct choice of tool depends on the situation.

From the point of view of understanding the customer's needs, especially within the context of a larger situation, one of the most powerful tools associated with Six Sigma is Quality Function Deployment (QFD).

Quality Function Deployment originated in Japan in the 1960s, and later on gained popularity due to success within the automotive industry. The Quality Function Deployment process utilizes a graphic called the House of Quality to identify customer desires and document their importance as seen in Figure 2-1. The figure has been simplified to some of the more complex real-life versions to show how correlations between the design inputs and correlations between design inputs and customer requirements can both be seen.

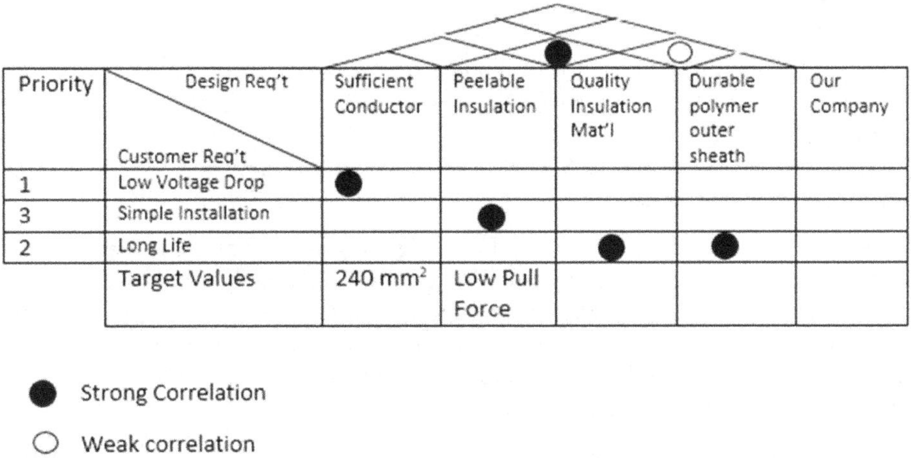

Priority	Design Req't ⟍ Customer Req't	Sufficient Conductor	Peelable Insulation	Quality Insulation Mat'l	Durable polymer outer sheath	Our Company
1	Low Voltage Drop	●				
3	Simple Installation		●			
2	Long Life			●	●	
	Target Values	240 mm²	Low Pull Force			

● Strong Correlation

○ Weak correlation

Figure 2-1. A simple House of Quality diagram, as typically used in Quality Functional Deployment. Note that the diagram shows correlations within the design requirements themselves and between the design requirements and the customer requirements.

Six Sigma also employs the Seven Management and Planning tools, which were popularized by the post-WWII Japanese approach of Total Quality Control.[12]

QFD is a method which allows its users to employ System Thinking and Psychology to their problem, which means they can develop a proper understanding of where the customer sees value. It covers both the "spoken" and the "unspoken" requirements to avoid the problem of developing a product that is precisely what the customer asked for, without being anything like what the customer wanted.

The key message is simply that what the customer values isn't necessarily what we think she values. It also isn't always the first thing the customer complains of when they initially engage. Discovering the real motivator for the customer can be difficult.

However, as doing so is crucial to selecting the right objectives, using the available tools to uncover the user's underlying motivations should be an essential part of the data scientists' process. Quality Function Deployment is a key example of the kind of tools that have been successful in understanding a customer's concerns in their wider business context.

---

[12]"What Are the Seven New Management and Planning Tools?", accessed April 7, 2019, from https://asq.org/quality-resources/new-management-planning-tools.

CRISP-DM

Six Sigma and the DMAIC process were not developed with data science or data mining in mind. Although we have recommended at least considering the use of some of the tools of Six Sigma to establish the customer's needs, an important way that the Six Sigma process is not suitable for a data science project is that it is linear.

The CRISP-DM[13] process allows for insights gained from the data to be re-incorporated into the understanding of the user's problem. The degree to which this will be suitable obviously depends on how available the user wants to make themselves to you; when deciding on an overall strategy for a particular project, it will be important to consider how available the user is prepared to make themselves for questioning.

The iterative nature of CRISP-DM—something it has in common with the Agile philosophy—makes it a good way to think about data science projects. On the other hand, as it lacks some of the customer focus, project management, and close out elements of other approaches, these may need to be borrowed from another approach.

The CRISP-DM cycle can be seen in Figure 2-2. By using the information gathered at subsequent cycles, a data scientist using the CRISP-DM cycle can improve on their first guesses. In particular, there is a specific provision made to go back to the customer for further discussion after the initial phases of data discovery.

Each of the phases is subdivided into smaller areas,[14] which contain checklists of important areas to consider. The Business Understanding phase, for example, being the phase that most corresponds to this chapter, has the distinct goals of understanding the business goals, as well as success criteria for the data mining project.

These are then referred to in the successive phases of the project. For example, in the evaluation phase, the evaluation is performed with respect to the original success criteria, as might be expected.

More so than Six Sigma, however, CRISP-DM is nonprescriptive, so no specific guidance is given around which tools that can help achieve the desired business understanding.

---

[13]*The IBM SPSS Modeler Guide to CRISP-DM,* ftp://public.dhe.ibm.com/software/analytics/spss/documentation/modeler/15.0/en/CRISP_DM.pdf.
[14]Ibid.

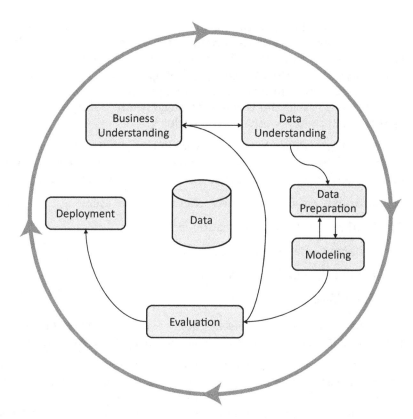

Figure 2-2. CRISP-DM cycle

Empathizing with the User

The approach taken by Six Sigma practitioners is very much an engineering approach. The key tactic is to define a concrete problem with a financial payoff. This is appropriate in the usual Six Sigma context where problems are most often associated with a specific cost of poor quality outcome—a need for rework or scrap, for example.

However, this may not be the best approach for every opportunity that can be tackled through data science. Where the opportunity is not tightly coupled with a specific poor quality outcome, a Design Thinking approach might be a viable alternative.

In a typical Design Thinking life cycle, developers build empathy with users before generating ideas that represent possible solutions. This approach allows the practitioners to get a better understanding of the customer's underlying or true problem, rather than the surface problem they may have presented with.

The process of empathizing with the user allows a wide range of possible solutions to be generated, with the potential for some very blue sky thinking. The elevator example from Mitroff and Silvers we saw earlier in the chapter in fact illustrates this perfectly—the psychologist thought about what the elevator users actually wanted rather than the kind of problems that she usually solved. As a result, she was able to suggest a solution that succeeded where a technical solution wasn't as successful.

There are a number of different ways of describing the design thinking process. One of them is the Stanford process, which uses a five-stage process to describe the steps need to ensure that the right problem is tackled.[15]

1.   **Empathize:** Gather information from your users.

2.   **Define:** Turn user information into insights.

3.   **Ideate:** Generate ideas based on insights.

4.   **Prototype:** Build a version of your ideas.

5.   **Test:** Validate your ideas.

Although these stages are written as a sequence, that doesn't mean they need to be performed in a strict sequence—iteration is possible and desirable.

The initial stages are particularly important in our context. Empathizing with the user means ceasing to impose your own ideas, and putting aside your own assumptions about what this problem looks like to the client or user. This is exactly what the psychologist did in the elevator problem. She asked herself, "As an elevator user, how am I affected by the time it takes for the elevator to arrive?"

By empathizing with the user, you will be able to understand the actual problem they are facing, rather than the problem that your own assumptions about the world make you impose on the client.

Don't Waste a Second

Time with clients is precious. Imitate experts in other professions—doctors, lawyers, management consultants—to apply the best interviewing techniques that ensure you get the most from the short time you have with your clients. If your clients are internal, you have more time, but it's still easy to waste it, and therefore to fail to get the best results—and it still isn't infinite.

---

[15]Rikke Dam and Teo Siang, "Five Stages in the Design Thinking Process," accessed April 8, 2019, from www.interaction-design.org/literature/article/5-stages-in-the-design-thinking-process.

Part of the problem is speaking the same language. There is no reason to assume that experts in other fields are conversant in machine learning and statistical techniques and jargon. The second difficult aspect is how to project confidence in your model or other data analysis without projecting arrogance.

Appearing, if not humble, at least in a way that means you don't appear to place yourself above the person you are trying to help will prevent turning them off. If you remember that the information they have about the nature of their problem is just as useful, while more difficult to obtain (there are lots of data scientists around the world, but only a handful of people understand your clients' problem like they do, and they are likely all at the same work site) should make it easy to stay grounded.

Getting a gauge on where your client's statistics or data analytics knowledge is also important. Explaining every basic statistics term to your clients when they already perfectly well what the median is will get them offside as surely as will baffling them with esoteric discussions of the spherical family of distributions.

In general, there is a strong chance that members of your audience have done a business degree. It's likely that they understand the basics of descriptive statistics. Concepts such as standard deviation may be fuzzy or misremembered, but they probably aren't completely novel.

Machine learning algorithms and their jargon are relatively rarely encountered by people who aren't trying to become experts in the field, the most obvious example being data scientists.

Consider carefully whether you need to tell an audience that your model may be based on a Random Forest. Before any models have been built, there shouldn't be any need to go into that sort of detail—for any model, the algorithm used to build it is one of its least important attributes. If you start getting side tracked into explaining how a particular algorithm works to a client who isn't going to build their own models, you will exhaust the time they have available to explain their business to you. Let's face it—there is almost no chance they have enough time to explain it to you properly.

Note   One of the catchphrases often heard in a design thinking environment is to avoid solutioning. Solutioning is where someone who is running a session with users or other local experts intended to define problems begins to suggest solutions. Doing this can be fatal to the session, as the local experts will stop explaining their problems and either shut down or offer their own solutions.

Better yet, hold back on the details of possible models until after you report on what you have built. Ultimately, you are not selling a Random Forest or a neural network, you are selling a reduction in their exposure to risk, or the means to reduce some of their costs. The customer isn't interested in how that is done beyond needing to feel confident that you really can do it.

In short, a discovery session with your client or user is a great time to remember the adage that "you have two ears and one mouth: use them in that proportion," and then talk even less. People often want to fill in silence with their own words—use it to your advantage. Think carefully about when you need to guide the conversation, and in what direction to guide it.

Conversations with the customer almost always turn up new views on what is really important. Sometimes all you need to do is sit in the room and listen. Other times you will need to coax the customer to the right frame of mind to tell you what they really need.

Data Science and Wicked Problems

There are multiple definitions of "wicked problem" but one characteristic that may be seen in problems presented to data scientists is that the lack of a definitive formulation.

Wicked problems are likely to be intractable if you try to apply traditional data science tools (some problems thought of as wicked problems have been at least partially solved using advanced forms of decision theory, but that's way beyond our scope). That doesn't mean that data science tools can't be used to indirectly tackle at least certain aspects of these problems.

It does, however, mean that is very important to identify when you are being asked to solve a wicked problem as early as possible. Once you know that that is what you are dealing with, you can decide an appropriate course of action. The two chief possible courses being pass on the problem, or restructure the problem into a tame problem.

Important markers of a wicked problem include the clients you talk to failing to agree on a common understanding of the problem. Attempts to define where the problem begins and ends may be fruitless. The problem may appear to be a symptom or a cause of another problem. In this way, the problem shifts its shape so as to defy attempts to place a boundary on it. This shape-shifting makes the use of any sort of model or algorithm extremely problematic.

As has been noted before, a significant part of solving a problem is defining it correctly. A common element of wicked problems is that they defy an easy definition. However, deciding on a stance beforehand sometimes forces the problem into a solvable state.

By reframing the problem in an acceptable way, it can become tractable. As data scientists, the preference is logically toward numerical solutions or solutions involving automation.

There are in fact a range of problem restructuring methods that are designed assist in transforming wicked problems into problems that can be easily solved, assuming the right inputs are available.

Strategic Assumption Surfacing and Testing was introduced in a paper by Mitroff (of the elevator problem mentioned earlier). The goal of this method is to understand the underlying assumptions that govern a problem.[16] This is done in groups over a five-step process.

Strategic Assumption Surfacing and Testing is just one of a range approaches to redefining problems so that intractable problems become tractable. There are now at least several methods with reasonably long track records to restructure problems.[17] While they have some common elements, such as usually being originally envisaged as a group activity (although they are often reworked for use by individuals), the differences between approaches mean that you can select the right one for your specific environment, and sometimes modify an approach with ideas from another approach.

Understanding the correct objective at the beginning of a project can be clouded if the problem has been set up as a wicked problem, and may seem intractable. However, as the form of any problem is usually determined by many assumptions—some easy to identify, some less easy—there is very often scope to restate the problem in a way that yields to your available tools.

Careful application problem structuring methods will often let you turn an ill-posed problem into a well-posed problem, and succeed where others have failed to gain traction.

Documenting the Project's Goals

Just as is the case for team goals, goals for projects can easily be forgotten or misunderstood. Strong documentation of the right information ensures that you don't fall at the last hurdle by delivering to the wrong target through a simple misunderstanding of what you have already discovered.

Aside from defining the target you are modeling there are multiple dimensions that may be of importance to the user. Obvious examples include ease of

---

[16]"Strategic Assumptions Surfacing and Testing," Institute for Manufacturing, the University of Cambridge, accessed April 7, 2019, from www.ifm.eng.cam.ac.uk/research/dstools/strategic-assumptions-surfacing-and-testing/.
[17]Jonathan Rosenhead, "What's the Problem? An Introduction to Problem Restructuring Methods," *Interfaces*, December 1996.

understanding the model's results, speed at which the result can be returned, and speed at which the solution can be implemented. It can be helpful to document these aspects also.

It is also helpful to document the data sources that are available to the project, and the platform that the results will be delivered to, to define the final format.

Individually these points of information may seem very trivial, but missing a few of them can at least waste time, if not lead to a project that fails to meet the client's expectations.

The notion of documentation has unfortunately developed an association with traditional waterfall project management paradigms, and the risk of giving a power-hungry project manager a blunt instrument he or she can use to beat other unfortunates over the head with. As noted elsewhere, this risk is caused by human personalities, not by the tools or processes themselves. The project manager who beats you over the head with a formal requirements document is more than capable of developing Agile rituals that waste everyone's time and lock the company up in pointless mummery.

It also pays to remember that documentation doesn't have to be formal. It's lovely to have comprehensive plans produced, which distill every brainstorm into their essential wisdom, but it's also very time-consuming.

Fortunately, we live in a time where documentation doesn't have to formal typed minutes in ring binders that no one will ever read or access. From wiki software, such as Confluence, through to virtual whiteboarding software, there is a range of options for capturing user discovery as it happens or for distilling the results of that discovery into documents with actionable insights.

The existence of electronic means to keep documentation is especially useful considering that things often change. When documentation was synonymous with hard copies, there was a psychological barrier to making those changes. Now there is no reason not to update documents as customer needs become clearer. In some cases this will require a change management process, if the documentation is formal and there is a risk of contradiction.

Other times it will be enough to collect the results of workshops as they occur as faithfully as possible and update them when necessary.

All in all, learning about the customer's real needs is a difficult process. It is time-consuming, and it depends heavily on the customer's own goodwill. With that in mind, we want to ensure that nothing is lost.

Selecting a correct objective and framing it to ensure the best results for a data science solution are essential elements to defining projects that will succeed and that will help your team maintain the credibility as expert problem solvers. However, although the correct objective may be the most crucial element, having the correct skills in each project team, and the best possible data set are also critical, and we will consider these elements in the following.

Ways and Means—Project Resources

In Chapter 1, we saw strategy presented as the equation "Strategy = goals + ways + means." In general, in a data science context, our ways and means are the people in our team and their skills, which represent the "ways," and the available data, which represents the "means."

Although in this chapter, we have focused on the risk of having a badly chosen goal, mostly because this is the least talked about risk, the risk of not having the right resources is also an ever-present danger, although one that is more likely to lead to a project not being finished, rather than an improper project being finished.

Either way, though, the risk is real to your reputation as a data scientist, and to the reputation of data science as a valid way to solve problems within your organization. To continue to maintain your license within your company to take on difficult challenges, you need to make sure that people see you succeed as often as possible—carefully considering to what degree particular projects are in your capability is vital.

We will consider the two prongs—first the capabilities that exist within your team—the ways. Second, the data at your disposal—the means.

The Ways—Data Science Skills

A challenge in this area that is bigger for data scientists than for other professionals is that the broad definition of data scientist sets up an expectation that any single data scientist has the skills of any other single data scientist. So far, the idea that there might be some data scientists who specialize in text mining while another who specializes in true big data hasn't penetrated fully, which makes it entirely plausible for the team to be asked to do something outside your expertise.

At the same time, the data science community to some degree buys into this idea by promoting relatively disparate areas of expertise, such as Deep Learning, Natural Language Processing, and Geostatistics as all equally a part of one skill set.

In this environment there is a risk with every project that it will be outside your capabilities. This, in itself, isn't necessarily a hard no. You should be taking on at least some projects that have been stretching your capabilities as one of their core purposes. The problems start when expectations haven't been reset to take into account the fact that the team's capability isn't quite there.

This is where the concept of the project contract, seen in systems like Prince2, is important—they allow you to agree to projects you don't have complete capability for, while explicitly setting the expectation that there is more risk attached to both finishing the project at all and the expected timeline.

When you want to stretch your capabilities, you need to identify projects that will fly below the radar, at least to some degree. Otherwise, your organization will continue to maintain expectations at the same level as for projects that are more clearly in your wheelhouse. If you can succeed at maintaining a steady trickle of capability stretching projects, however, you will be able to develop skills.

As long as you are clear that the capability to achieve the full set of expected benefits may not yet exist, completing projects that are a little outside—or sometimes a long way outside—of your capabilities is a good way to improve capabilities. The important thing is to ensure that everyone is clear that stretching capabilities is actually a more important goal for the project than the nominal project goal.

The Means—Available Data

The final part of the ways and means equation as it applies to data science is the means—data. You obviously won't always have all the data you need to model anything your customer wants.

Both Agile and CRISP-DM offer a partial solution to this problem. In the CRISP-DM cycle, the next stage after Business Understanding is Data Understanding. By arranging with the Project Sponsors or Clients to make the Data Understanding stage a toll gate, you can manage expectations around the possibility that the data isn't plentiful enough or of sufficient quality to support the goals of the project. Note too, that, in the CRISP-DM process, there is room at this point to iterate on the goal of the project—to nudge the goal into something that is more achievable (the essence of strategy, in a sense).

An Agile approach allows something similar by making an initial assessment of the data an early deliverable that sets the scene for future deliverables. Hence, similar to the way we took each stage as a deliverable within the CRISP-DM framework, we can do something similar within Agile. We would define an Exploratory Data Analysis report as a deliverable, and then a prototype of the model as a second deliverable.

At either point we have an option to either end the project, obtain different data, or pursue a different goal at each stage. If we're happy with the prototype, the implementation can be its own set of deliverables, each time becoming more complete. At any time, based on the results of model evaluation, you may decide to build in a stage of capturing or cleaning additional data.

Always remember that data is elastic. That is, when you need more data, sometimes more data can either be obtained (i.e., bought or otherwise acquired from someone who has the data now) or gathered (i.e., a sensor or procedure can be put in place to trap the data). Either way, time or money (which can be seen as having a similar equivalence to energy and matter) is frequently the only thing standing between you and more data, so if your case is persuasive enough, your organization will often get it for you.

When you are considering which data you are going to incorporate into your model, consider whether the data that you score the model on will be at the same quality as the data you are building the model on. It's common to use a historical data set to train the model, to begin with, and as that data has had time to be considered and reconsidered, late arriving data has had time to arrive, and problems with the data have had time to be corrected.

In contrast, the scoring data will be much closer to live or real time. As a result, there may be more data missing, and more of the data may be incorrect or poor quality in some other way. Keep an eye for these sort of problems when you intend to recommend a model for implementation.

Data is the vital clay for any data science project, so assessing its availability early in the project life cycle is crucial to ensuring that your project will achieve the desired results.

However, the quantity and quality of data are not carved in stone, and additional data can often be found or gathered without incurring too large an expense. Build the right case, and the data will come.

The Project Hopper

There is always more work to be done than anyone can do. There are more projects to do than anyone can do also, but something that isn't worth doing now might be the most important thing in the future.

A data scientist who works within an organization can benefit from the use of a project hopper—sometimes seen in Six Sigma organizations. The idea is simply that sometimes when you have completed the Define phase or the Business Understanding phase, the decision will be made that there are other activities to work on that are higher priority. However, it is obviously wasteful to simply dump the work that has already been completed.

Instead, summarize the work that has been done so far and create a project hopper which keeps at least the outline of the business understanding section for later.

The hopper is an especially good place to store projects that have increasing skills as one of their primary goals. You can record the skills you hope to gain by doing the project, so that you can develop new skills for other projects likely to attract more scrutiny on a just in time basis.

For example, if you can see a high-profile project on the horizon that is likely to require deep learning, you can select a lower profile deep learning-oriented project from the hopper to ensure the required skills are up to scratch.

The unicorn data scientist, who understood every facet of statistics, machine learning, and programming, is increasingly recognized as the mythical creature she always was. Validating whether the skills to perform vital projects exist within your team members needs to be a priority.

Ensuring that you maintain a steady flow of projects that have a key objective of stretching your team's skills will ensure the right skills exist when you need them. At the same time, ensuring the rest of the organization understands which activities are within your team's wheelhouse means that expectations remain realistic about what you are able to achieve.

The project hopper is a great strategic tool that allows the manager of a data science team to take control of the flow of their team's work on the one hand, and on the skill building and overall direction of the team on the other. By using it properly, you will be able to successfully build new skills for your team on projects that aren't at the center of your organization's attention, while also completing a steady stream of projects that meet your organization's most crucial goals.

The project hopper is also a place we can keep excellent projects that are missing one of the three elements—projects with clear objectives but no data, projects with great objectives and a viable data set but requiring skills not yet found in the team, to name two examples—but will make great projects in the future.

Summary

Project risk is often considered from the point of view of risks to timely project completion of the project only. Less often, project managers and their project sponsors consider properly the risk of completing a project that doesn't solve the client's problem or one that creates a new problem.

Sometimes creating a new problem moves into the territory of "unknown unknowns," as they were termed by Donald Rumsfeld. We can't always prevent this happening, but we can make it less likely by carefully considering the voice of the customer.

There are a number of ways of doing just this. The Six Sigma DMAIC process emphasizes the voice of the customer in the initial Define phase and proposes some tools for better learning the voice of the customer that are applicable in some data science contexts. One of the most powerful tools available within Six Sigma is Quality Function Deployment, which employs Seven Management Tools within the House of Quality framework to expose.

CRISP-DM, which has been designed as a standard framework created specifically for data mining, describes an iterative process between Business Understanding and Data Understanding. This has the advantage of allowing the data scientist to refine their Business Understanding by reference to the available data—presenting data findings aids the conversation with the client.

Finally, the Design Thinking approach that is associated with Agile promotes empathizing with the user and suggests another set of ways to get closer to the customer's expectations.

All of these require the best use of the client's time when you are able to spend time with them. As much as anything else, they are unlikely to have an unlimited amount of time to give you. Hence, we cover some techniques for ensuring this time is put to best use.

Data science has a reputation as being able to solve the most difficult problems available. This reputation is a double-edged sword, as while on the one hand it means that data scientists are given the chance to work with the most challenging and interesting problems, on the other hand they are given a lot of license to develop their preferred solutions.

However, the flipside of this opportunity is that sometimes the problems given are genuinely unsolvable. A notorious category of unsolvable problems is called wicked problems. Identifying them gives you an opportunity to turn them down, and therefore having your reputation by a problem that was unsolvable from the outset. Alternatively, you can attempt to convince your client to allow you to reframe the problem into a format that allows it to be solved. A number of techniques exist to do just this.

Although well-chosen objectives are vital to the success of a data science effort, and moreover to the perception of success, they aren't the only important aspects of a successful data science project. In the previous chapter, we also discussed how ways and means—the skill set within a project team and the available data—are also vitally important to the overall success of a strategy.

In the context of data science, the ways are effectively the team's skill set, and the means is the available data. At a project level, careful attention needs to be paid to whether the team has sufficient skills to complete a project. This may sometimes lead you to turn down a project.

At the same time, trying projects outside your current skill set is the best way to develop new skills. With this in mind, sometimes you will want to attempt projects clearly outside your current skill set. When taking on such a project, it is important to ensure that expectations within your organization are properly managed with respect to the likely timeline, and the probable efficacy of the final product.

In this chapter and the previous chapter, we've looked at how to apply strategic thinking at a data science team level, and at a data science project level. In both cases, the goal was to ensure that your efforts as a data scientist achieved their objectives and were fully appreciated.

In Chapter 3, we will look at how to sell data science teams and data science projects, so that you are able to maximize the usage, and therefore usefulness of the projects you work on. More fundamental than that, being able to sell the projects you are working on means they will get a green light to begin with.

Project Checklist

This checklist with items worth considering while during projects is divided into three parts—objectives, skills, and data.

Objectives

- Are there regulatory requirements, such as usually the case with financial or insurance models? What is their effect, for example, do they restrict algorithm choice, require additional documentation, or additional reporting during the model's life?

- How often will the model be updated? Possible answers could range from "never" down to every micro-second.

- What are the consequences when (not if!) the model is incorrect? Nothing? Someone loses some money? Someone loses their life (could be the case, for example, for a medical diagnosis model)?

- What volume of data will be fed to the model? What turnaround time is acceptable for the model's results?

- How will users access the results?

- How much access will the data science team have to end users? Will the data science team be able to access the end user more than once during the project life cycle?

Skills

- Do the skills exist currently within the team?

- Are the people with the right skills also the people who are available?

- What will the consequences be if the project is not completed?

- Is the project urgent?

- How difficult would it be to hire a temp? How much of a delay would hiring a temp cause to the project?

Data

- Has the team worked with this data set before?

- What is the provenance of the data? How likely do you believe it is that the data set is of good quality before you begin to explore it?

- Would you get a better result with more data? How much would it cost to gather more data? How long would it take?

- If you incorporate this data into your model, will you have permission to use the data when the model is implemented?

- When you implement the model, will the data be refreshed as often as you need the model to be refreshed?

Data Science Sales Technique

Getting Your Project Adopted

So far, we have seen how to align our objectives with the larger organizations and how to develop a strategy that plays to our strengths. On a day-to-day level, however, we can't assume that people will adopt our ideas or implement our projects—even if they asked for them! They also need to sometimes be reminded that we are there, and that data science is there also, ready to solve their problems.

Selling Data Science Projects and Ideas

The art of selling appears to be often overlooked both by data science curricula and by the myriads of bloggers and authors that write enticing articles for emerging data scientists. Often, when reading about data science, you can get the impression that building a model with a great accuracy score and then constructing a data pipeline to service it are the start and finish of the job.

The reality is that to be allowed to build and then implement a model, you will have most likely needed to sell yourself, your idea for a model, and then your actual model to people at different stage gates along the way, and frequently there were different groups of people at each of those stage gates.

© Robert de Graaf 2019
R. de Graaf, *Managing Your Data Science Projects*,
https://doi.org/10.1007/978-1-4842-4907-9_3

Often the task of getting past those gatekeepers is reduced by data science writers to a matter of storytelling, and it is true that effective storytelling can be key to success. However, there is a lot more to convincing stakeholders that you are equipped to discover a solution to their problem, and then convincing that your proposal is that solution and should be implemented than telling the basic story of your data analysis.

In this chapter you will learn how to tailor your sales techniques to your customers, how to sell your model, how to sell your team, and how to sell data science as a versatile solution to your organization's problems. I will also look at how your pitches should vary at different stages of your project, and how to make storytelling work to sell your project, instead of just explaining your data.

Beyond that, we will discuss the need to market the art of data science as well as yourself to ensure that your solutions are seen as the key to solving your client or your organization's problems.

A key thread running through this entire chapter will be the need to go beyond your purely rational mathematical arguments into considering the emotional needs of the people you are trying to convince. Although this may be challenging, it will enable you to take your arguments a long way further, and therefore make the best use of the carefully crafted rational arguments that you may be more used to.

Mastering the emotional side of persuasion, in conjunction with the rational side, will give you a strong advantage when it comes to getting your models adopted.

Selling Your Data Science Project

No matter how great your data science project is, it's all for nothing if no one out there wants it. If we can't convince other people that our work will improve their lives, or at least has that potential, it will just sit on a server somewhere until bit rot stops it functioning.

But what's the right way to go about telling the customer why our work is going to change their lives for the better? Creating sales documents or presentations that list out all the shiny new things that our data science application can do is very tempting. We worked hard on those features and everyone will appreciate them, right?

Well, not really. For one, it's very likely your target audience doesn't have the technical ability to understand the point of what you're selling. After all, if they had your technical skills, they wouldn't be thinking of hiring a data science, they'd just be doing it themselves. When you're creating a sales document, the first thing you have to is reduce the references to the latest tools to the barest minimum that lets them know you know what you're on about.

The next problem is that you can't trust that the customer realizes how your solution helps them out of their present predicament. Moreover, it's disrespectful to get them to do your job for you. Hence, you need to make sure your pitch joins the dots between what you intend to do for the customer and how it's going to make their life easier.

I will now focus on what kind of model to present to a customer at a pre-implementation meeting where you're looking for the customer to say yes to implementation and therefore commit more time and money.

This stage could turn out to be the most difficult hurdle to get over because this is where the time and money commitment increases from a small amount to potentially a much larger commitment. In an internal sales situation, this could mean the executive is deciding whether to move the project from within a small data science to a larger team who will need to implement the project and allow it to be used widely through the organization.

Obviously in this scenario, it is vital to be able to link implementation of your model to solving a real-world problem your customer is experiencing, and that the real-world problem should be important to the customer. Where possible, you should be able to link solving the particular problem with saving a certain amount of money.

You will be more likely to be able to find a reasonable dollar value when you are presenting a model internally. At the same time, if you have been successful with your earlier discovery meetings, you will have some sense of how important the problem you are working on is to your customer.

The key to buying model credibility is that the results make sense to the customer. That means that not only will your model need to show great results but also your customer will need to understand how you evaluated your model, relate your evaluation to her business, and believe in the results.

There are many available evaluation methods, and their use can provoke controversy within the statistical and data mining communities. Frequently used evaluation methods such as the receiver operating characteristic have been criticized on statistical grounds, with more robust but nuanced (and therefore more difficult to understand) alternatives proposed, and yet even these simple but flawed methods can be hard for a business audience to understand.

There are also methods such as lift or gain that are tied tightly to the business problem that the customer is trying to solve. In the context of a presentation to sell your results, this type of evaluation is ideal, where the data and problem are suitable. For example, lift is explicitly tied to the marketing goal of increasing sales.

If these metrics are not suitable, developing your own metric that fits the problem may be a way forward. In either case, it is still best practice to also perform the evaluation using a statistically robust method to ensure a correct assessment of the model's performance without necessarily using that method to communicate the results.

Ultimately, you need to make sure that you don't let the perfect be the enemy of the good, and when you are communicating results to customers and clients choose the evaluation method that will be understood and accepted by your audience.

Now, although for a few shining minutes it was okay to produce a model that was very accurate without knowing how it came to its conclusions. Certainly, you can still win Kaggle with a model of this kind. Unfortunately, though, it can be very difficult to convince a user to trust a model with an accuracy score alone. On the one hand, many metrics used to evaluate models are inaccessible to people, who themselves are not machine learning users. On the other, for many people an accuracy score isn't convincing on its own behalf—such people wonder, "Will it work with next tranche of data?" and "Is it too good to be true?"

Self-Promotion

Many people in the data science community are better at doing data science than they are at marketing themselves. This is only natural—if marketing and promotion came naturally, they'd probably do that job instead.

However, as we have seen, within an organization, it is usually not enough to do good work. It is important that your good work is seen and recognized. Beyond what you have already achieved, you also need to ensure that your organization can see what you can achieve, or what you could achieve, if only you were given challenging enough projects to work on.

You need to do this both as an individual and on behalf of your team, whether you are the team manager or not. In both cases, a big chunk of what you need to do is essentially managing upwards. This is obvious in the case of the team member but is true for the manager as well because it is often the case that (excepting specialist data science businesses) the data science function is a relatively small department feeding into something larger.

Consequently, the team leader will need to devote a substantial amount of time to explaining the value of their department to the business at large, including both other team leaders and their teams and more senior managers.

You can find a lot of advice on the best way to manage up with only a cursory search in your library's management books section or on the net. Obviously, some of it is more applicable to the situation we are looking at than the other.

For reasons of space, we will just look at a few of the most easily applied advice.

The first piece of advice is to communicate. Similarly, to the individual team members we will discuss below, you need to keep your senior managers up-to-date with your achievements. Even more than you don't want to hear about every problem your team encounters, senior managers are even less likely to be interested in the difficulties you had along the way to what you've achieved. In fact, beyond a sketch, they won't be interested in the how at all. What they're mostly interested in is the business problem your innovation will remove, and how much money they will need to invest.

A lot of the persuasion you are going to need to do to get people's trust who are either above you or sideways at the same level will need to be indirect persuasion. People in those positions who don't report to you, and who sometimes may feel threatened if you are successful, won't always respond to your direct message, even if what you are trying to explain is logically in everyone's best interests.

If you are the team manager, individual successes by team members will have a habit of being seen as your individual successes. You may feel that you deserve this credit, at least inasmuch as you probably assigned the task to someone initially, gave them the tools to do, and prepared a smooth path for them to achieve their results.

At the same time, you need to take enough of the credit for the team's successes to demonstrate that you are an effective leader. However, in the long term, this will cost you as the team members who are achieving on your behalf will feel cheated out of praise that more rightfully belongs to them.

As an individual, you have a little more freedom to take credit for your work, given it will often be credited back to the team as a whole. However, there are two common mistakes that I often see. One is presenting work too early, and the other is not presenting it, or holding it back for too long.

If you are a data science team member, you need to keep your team leader up-to-date with your progress. However, she doesn't necessarily need to see every twist in the path. It's reasonable to update her when you've either reached a point she will recognize as a milestone or when you've hit a genuine block.

In both cases, you need to put yourself in your team leader's shoes to a degree to know whether you're really there. What would someone on the outside of your task understand as a step forward? What would your team leader expect you to have tried first before asking for her help? At a minimum, she would probably expect you have looked for help in a few of the "usual places"—your other team members, the documentation, StackOverflow, etc.

It's also important that you consider the best time to present your idea. Buttonholing your manager just before close on a Friday won't be such a good idea. Ideally, give her some warning that you want to update her on your progress—even if you have an open plan office and she sits right next to you. That way, you'll have the best chance that she will be ready to talk and think about your work at the same time that you are.

Ensuring that some credit for success flows back to the team is important because it drives the rest of the business to come to you when they have a problem that needs solving—and it drives them to come to you early, before premature "solutioning" means that you need to attempt an inappropriate solution before being allowed to solve the problem properly.

In exactly the same way that it is important for team members to consider how to communicate their achievements upwards to their team leader, it is just as important for the team leader to create opportunities for her team members to showcase their work.

Tailor Your Message

Even when you are talking to people from the same organization, different people performing different job roles have different concerns. Sometimes, when people change jobs, you can even witness the change in how they view different problems.

At the same time, it will be common to present to groups of people, often with conflicting concerns. It can be very damaging to give a presentation that addresses none of a particular stakeholder's concerns, especially if that person is especially influential in the organization.

Williams and Miller, writing in the Harvard Business Review,[1] outline five different types of decision makers, who each need a different approach if you are to convince them your idea is sound. Their listing is as follows:

1. The charismatics

2. The thinkers

3. The skeptics

4. The followers

5. The controllers

Each of these requires a tailored approach and differing preparation. The overall message here is that you need to understand who you are talking to,

---

[1] Gary A. Williams and Robert B. Miller, "Change the Way You Persuade," *Harvard Business Review*, May 2002, https://hbr.org/2002/05/change-the-way-you-persuade.

and the best way to persuade that person. Sometimes you will even need to consider more than one different person in a presentation audience, and ensure that your presentation speaks to each of them.

However, knowing these aspects of your audience won't be the whole of your task, and awareness of the existence of these audience types isn't enough by itself to ensure success. For that you will need an overall guide to persuading others.

Conger,[2] also in the Harvard Business Review, breaks down the elements of making your case persuade people when your position doesn't give you authority over them.

I have adapted Conger's tips for data scientists.

1. Establish credibility. This should be the easy one for a data scientist, and it is true that the job title "data scientist" gives you instant street cred as the local smart person. However, what is often missing is credibility that you understand the business, and, more importantly, that you understand what others in the business are going through. The secret of success here will be leaving aside your data science expertise, and focusing on your knowledge of the business and the immediate problem.

2. Frame goals on common ground. Similar to establishing credibility, this depends on applying your knowledge of the business and your audience to ensure that the solution you propose will meet both the business' needs and their needs.

3. Emotional connection. In some ways the most difficult, if you have developed the habit of using facts and reason to persuade. This is a two-way connection, where you can express your own emotional commitment to an idea, but also where you are sensitive to your audience's emotions. The latter is the more useful of the two—learning to read your room's emotional state will enable you to correctly calibrate your presentation.

Conger also proposes four ways of approaching persuasion that are best avoided:

1. Trying to make your point with an upfront hard sell

2. Resisting compromise

---

[2]Jay A. Conger, "The Necessary Art of Persuasion," *Harvard Business Review*, May-June 1998, https://hbr.org/1998/05/the-necessary-art-of-persuasion.

3. Believing the art of persuasion consists of making a great argument (or arguments)

4. Assuming that persuasion is a one-shot process

Of these, the last two are my picks for the ones most likely to be a problem for data scientists. In the first case, a data scientist is likely to be a person who puts a great store in the content of their argument, rather than the way it is packaged, in some ways as befits someone who has a pretense to apply a scientific approach. Sometimes it can almost seem that spending time on how an idea is packaged is a little shameful—making an idea look better than it really is could be seen as a little like confidence trickery.

Unfortunately, how an idea is received is only rarely related to its intrinsic quality. This can be easily seen from some of the ideas corporations of all sizes take to market—organizations adopt many bad ideas, so it is almost certain that they reject many very good ideas.

Rather than being won on the soundness of their logic, it is far more common for ideas to win the day when they are able to make a strong emotional appeal to their prospective audience. While this goes against the grain for many data scientists, what you should consider is that the idea that you are sure is the superior logical idea will lose out to another path that is logically inferior unless the right emotional appeals are put to the decision makers. For this reason, you need to take the need to frame your idea in the correct emotional terms seriously.

One way of looking at the emotional side of the argument is that it could be seen as the last mile of your persuasion effort. Of course, the rational side of the argument must be sound, but attention to the emotional side of the argument is needed to ensure that your audience or client doesn't refuse to engage with your argument rationally. We will see more of this later in the chapter when we talk about the role of emotions in winning trust.

Given that the emotional aspects of persuasion are there to ensure the success of rational argument, your appeal to emotions needs to be done without undermining your reason-based case. From that perspective, a little can go a very long way, just like salt or spices in cooking.

You can add a touch of emotion to your argument by doing things like using a rhetorical device, in the manner of a political speaker. Also try to break down barriers between the audience and yourself by personalizing your speech. Opening a presentation with a personal story that turns out to be relevant to your overall message is a good way of narrowing the gap.

Overall, you need to aim to create as much connection between the audience and yourself, in order to build the idea that you are all in it together. That way, the audience will stop thinking of you as someone who is trying to sell to them, and start to think of you and them as part of the same team, who are trying to achieve the same goals.

Benefit Sales

In sales jargon "selling the benefits" is making it clear to the potential customer how buying your product will improve their lives, and has been encapsulated in the phrase "nobody wants to buy a bed—they want a good night's sleep." The rub is that in most data science scenarios the problem that corresponds to the potential benefit is a business problem—such as reduced inventory or decreased cost of sales—rather than a human problem, such as a getting a good night's sleep.

Therefore, being able to complete the journey from feature to benefit requires some knowledge of your customer's business (whereas everyone knows the benefits of a good night's sleep—and the horrors of not getting one—far fewer under the fine points of mattress springing and bed construction) and the ability to explain the links. This last is crucial, as the benefits of your work are too important to allow your customer an opportunity to miss them.

What all this means in the end is that the approach of inspecting data sets in the hope of finding "insights" will often fail, and may border on being dangerous. Instead, you need to start with what your customer is trying to achieve, what problems they are facing before seeing which problems correspond with data that can be used to build tools that can overcome the problem. In this area the old adage that "you were born with two ears and one mouth so listen twice as much as you speak" comes into play.

If you became a data scientist in part because your temperament was better suited to quietly analyzing data rather than glad-handing customers, the whole notion of selling may sound daunting. It shouldn't sound too daunting, though. This advice works equally well for internal sales—selling to other areas of your company—as it does for external sales. If you're trying to sell internally, there should be many opportunities to find out what causes other people in your company pain. People love talking about themselves, and many people in the world can't pass up an opportunity to complain.

If external sales are involved there's a good chance a team sales approach is being used or could be. In this case you will be there as the technical support to a lead sales person (or people). Let your lead sales person do what they do best—cultivate the relationship, get the lead, introduce your business, and get the customer excited.

Use the time to learn as much as possible, flush out where the customer's life isn't as easy as it could be, and match your analytic solutions to making it easy. If a customer's life is painful enough, the remotest prospect of someone fixing it will have them reaching for their checkbook in no time.

For management consultants, an interview guide is an invaluable tool to get the best out of time spent with clients. However, it could be a little unsubtle and strange to use one with an internal customer, when you are trying to maintain an ongoing relationship with them.

That doesn't mean that going through the process of preparing an interview guide won't help you. Just the act of writing it out will mean that most of your questions are at your fingertips, as the process of developing the guide will force you to think carefully about the questions you are going to ask in the interview.

Closing

There is a lot of advice for salespersons on the art of closing the deal. A lot of it is too high pressured for data science sales where the targets are people who want to be sold to. Tactics favored by door-to-door salesmen are likely to have the wrong effect; in any case, if final persuasion is actually required with an external customer, this will likely be handled by a sales professional from your company.

More important from the point of view of a data scientist is that the details of what is to be done are as nailed as possible. Often there will have been a long lead time, where possibilities were canvassed. During the initial phases, this is often necessary, to convince a potential client that all problems can be solved. At the point when they sign on, however, it's now important to narrow the focus considerably, to one solvable problem, which is also defined narrowly enough that a solution can be provided that is recognizably what was promised in the contract.

This has a double benefit, in that the smaller-sized project will be less costly in both time and money. Moreover, if you are doing work for another company, it's likely that the people you are trying to convince will need to convince their own management, and a small, targeted effort will be more likely to succeed.

Promoting Data Science

Data science as a discipline has been enjoying a privileged position as the "in thing" for a few years now. This is a blessing and a curse. It is a blessing that people with the expectation that their problems can be solved or alleviated with data science are more likely to put their trust in data science solutions. It is a curse that people expect data science to solve their problems at the touch of a button regardless of whether there is reason to think that there is a good fit.

There is also an ironic aspect to this that there will be some people for whom the very fact that there are such glowing testimonies will cause them to be all the more skeptical. Maybe they are right—in recent times, data science has been near the peak of its hype cycle, and its intrinsic to a hype cycle that the activity or phenomenon being hyped receives more praise than its due.

The problem is that the hype becomes noise that obscures the most useful "real" part of whatever being hyped's benefits. So many fanciful claims get made that they obscure the sensible and practical claims. Therefore, there is as much a need for careful promotion when something is already overly hyped as there is when it isn't.

Careful promotion, however, is something that is clearly very different from hype. Rather than talking up pie in the sky claims for the way that data science and artificial intelligence (AI) could change the world, careful promotion means putting the case that data science is a useful tool when applied to the right problem. This means being realistic about the occasions when data science is not the right tool as well as extolling it when it is.

Anyone you meet by now will have heard about the way that the machine learning and AI is going to change the world, and many of them will have made up their mind whether they think what they have heard is either complete twaddle or a prophecy that is as important to humankind as the second coming.

The Double-Edged Sword of the Hype Cycle

Everyone has heard quotes about data science. "Statistics is the sexiest job of the 21st century," and all that. The hype cycle for data science has been surprisingly long-lived, and in 2018, Gartner[3] wrote "the hype around data science and machine learning continues to defy gravity."

Dealing with hype can be a problem in many walks of life. An article in the LA Times[4] describes the difficulties that young athletes face when they receive too much praise early in their career. In fact, this was part of the difficulty faced by Billy Beane of Moneyball fame that has been linked to his not living up to his potential when he was a player.[5]

The hype in data science doesn't occur at an individual level in the same way. When people in data science are singled out as particular rock stars, it tends to be after they have established track records. The problem for data scientists is more that the reputation of an entire industry precedes them.

There are two contradictory symptoms when your audience has been exposed to too much data science hype. The first is a jaded response to anything you say; the second is an audience member with inflated expectations.

---

[3]Hare Krensky, *Hype Cycle for Data Science and Machine Learning, 2018,* Gartner.
[4]Clay Fowler, "Dealing with the hype on social media can be a boon or burden for young athletes," Daily News, January 31, 2016, www.dailynews.com/2016/01/31/dealing-with-the-hype-on-social-media-can-be-a-boon-or-burden-for-young-athletes/.
[5]Michael Lewis, *Moneyball* (New York: W.W. Norton and Co., 2003), p. 3–13.

The remedy is the same in each case—reframe the discussion somewhere new to force them to forget what they've already heard and reconsider the topic from scratch.

Hence, when discussing a business problem that might be solved using data science, the key is to look at the problem from a different angle than the angle most often used in the hyped approaches to data science. A particular angle to avoid is "the power of big data" or the idea that effectively a big enough data set renders other considerations meaningless.

Instead, draw attention away from the kind of big sky thinking that's stereotypical of machine learning hype from the concrete gains that can be made within your organization. Avoid trying to find parallels with the large tech companies, such as Google or Uber.

Instead focus on moderate and realistically reproducible gains made by smaller-sized companies. Find use cases from within your audience's industry, and stick with those examples. Ensure that the use cases you pick are relevant to the business problems faced by your audience and have gains measurable in dollars, increased sales, reduced effort, or a similar metric that is relevant and understandable for a business-oriented audience.

If you are stuck for finding such examples, a book such as Eric Siegel's Predictive Analytics[6] which promotes data science is a useful resource. It's the kind of book whose only mission is to tell the world that data science is the best thing since the transistor. As a data scientist already, you might, therefore, not need this kind of book. However, the real-life examples are a great resource when planning a talk or pitch, especially when talking to an audience in a less familiar industry.

While there are numerous entries in the "data science is awesome" category, Siegel's offering is notable for its case studies over a wide range of industries—a special section in the first edition boasts 147 examples of predictive analytics applications.

By sticking to this plan, you will be sure, at a minimum, to give the audience what they are really looking for. However, if you are successful at providing a relatable example of data science in action, you will be able to make the audience member who thinks about data science in an unhelpful way reconsider how they think about the benefits of data science in their situation.

The best way to approach people who are tired of the hype is to bring things back to basics. You can demonstrate that you can be trusted—that you aren't one of the snake oil salespeople—by talking about data science in measured terms to explain its real benefits.

---

[6]Eric Siegel, *Predictive Analytics* (Hoboken, NJ: Wiley, 2016).

Branding—Personal and Shared

Arguably, data science has evolved from being a descriptor or classifier for a particular occupation, and become a brand. Of course, any label for an occupation becomes a brand at some point. Lawyer and medical doctor are terms that in addition to describing the kind of work their bearer does also evoke various stereotypes about what kind of person would do that kind of work.

Data scientist, as a label, has been around long enough to do the same thing. However, "data science" doesn't have the advantage of a professional body such as lawyers, accountant, or engineers have. Apart from setting professional standards and facilitating networking, the professional bodies for each of these professions have also taken on the task of marketing their professions.

As the "sexy" profession of the 21st century, data science has an advantage over other professions. However, these advantages are more obvious when it comes to attracting new entrants to the profession than they are when it comes to persuading people to listen to data scientists' advice. In recent years, we have seen professions such as accounting and actuaries build marketing campaigns to expand their professions' reach in relation to the areas that they can provide value.

In the latter context, the care that the professional bodies of professions such as actuary are putting into building their brand has the potential to reclaim some ground lost to data science, both in the context of recruitment and the context of users taking actuarial advice.

With no central body, these kind of campaigns don't occur for data scientists, who then have little control over their profession's image, which is often conflated with AI and big tech, meaning people with data science who don't fit that mold may find they struggle to meet people's "expected" notion of what a data scientist is.

To counter this, individual data scientists need to both be ambassadors for the data science brand, and to pay careful attention to their own personal branding. In each case, you need to have a strong concept of the image you are trying to project.

The oft-quoted idea that a data scientist is "someone who codes better than most statisticians, while knowing more statistics than most programmers" is unhelpful for this purpose, as it doesn't relate to how a data scientist can be useful.

The occupational branding used by the accountants and actuaries reflects their value proposition a lot better. The equivalent for data scientists might be that a data scientist will reveal what your data is telling you, or a data scientist is someone enables you to get the best value from their data.

Based on these labels for what a data scientist is, there is a grain of truth in the "programmer/statistician hybrid" conception of a data scientist. A data scientist is someone who straddles the divide between two worlds, however, the programmer/statistician divide isn't the important divide. Instead, the important divide is the divide between business and technical, and a data scientist is one of the groups of people who straddle that divide.

Once you have established your own definition of what a data scientist is and worked out how to market that definition within your organization and your wider network, you can extend and deepen that definition into your personal brand.

You are both an example of a data scientist, yet also other things besides being a data scientist, so tighten your data science definition to establish what kind of data scientist you are while at the same time defining a brand for yourself that includes your non-data science attributes.

This means that the relative lack of a professional body pushing a standard concept of what a data scientist is, and why they are useful is a double-edged sword. Professionals who have a professional body determining the brand are less able to decide their professional branding to suit their own strengths and weaknesses in the same way.

Moreover, older professions, such as accountants and actuaries have found that half the branding battle is to fight against a preconceived idea that doesn't suit them—in fact, both of those professions have had to fight against a stereotype of being boring.

Having decided on a definition of data science that plays to your strengths, you can move to building your personal brand around it. While a lot of data scientists might be the kind of people who would like the quality of their work to speak for them,[7] the reality is that you can't leave others impression of you to chance.

Don't Shoot Yourself in the Foot

Before we plunge too deeply into this topic, I'd like to mention that if you do a Google search on "personal branding," a lot of what you will read will be advice around your personal digital marketing strategy. That is, they talk about how to improve your LinkedIn profile, or how to use Twitter effectively to get a better job. These are important issues from one angle, but the kind of personal branding I will discuss revolves more closely around the way you are seen by your co-workers and clients and therefore has more to do with how they are able to see you act.

---

[7]Joseph Liu, "5 Ways to Build Your Personal Brand At Work," *Forbes*, April 30, 2018, www.forbes.com/sites/josephliu/2018/04/30/personal-brand-work/#63d19da87232.

As the previous paragraph makes clear, your personal brand is established whenever others encounter you directly or indirectly. In contemporary life, that often means via social media, but it wasn't all that long ago that individuals who weren't celebrities mostly encountered each other either directly or at one or two removes via someone else who had encountered the individual directly. Hence, you built your personal brand mostly by the image you projected to people who were in the same room as you.

This is still true within your own organization, where people see you in meetings, in the tea room, and at your desk with great regularity. What does what people see in these interactions say about you? Does the way that you arrange your desk project the image you are trying to cultivate? Co-workers, your boss, and your boss's boss all see the way that you keep your desk—is the impression you make, the impression you want to make.

On the other hand, perhaps your boss has access to an executive kitchen, so she never encounters the dirty dishes you leave in the sink. Even so, there are still plenty of opportunities for you to damage your brand in your interactions with her.

This stuff can seem trivial, and in a sense it is, but it's the prerequisites for building your personal brand. If you don't get those parts right, all people will remember is that you're a slob. Having set down those foundations, you can move on to establishing yourself as a trusted advisor within your organization. All in all, it's hard to get people to take you seriously, so don't lose the battle before you begin by failing something obvious.

Opt Out of the Boxing Game

A natural tendency for people meeting for the first time is to define themselves by what they do for a living. We often answer the question "What do you do?" by stating their job title or occupation name. However, for most possible audiences, a bald statement of "I'm a lawyer" will do a poor job of explaining what it is the speaker does to earn a crust.

On the one hand, most people are aware of a difference between a defense attorney and a corporate litigator, and many are probably aware there are lawyers who spend a lot of time in court and others who've never set foot in one. At the same time, most people don't have a strong idea of what a lawyer does all day.

Worse than that, when you answer "I'm a lawyer," the person asking puts you in a box, and when you ask the same question and hear "I'm an accountant," you do exactly the same thing. This is a boxing game.

In the case of data scientists, your audience will lean toward the "no idea what you do all day" end. Answering the question "what do you for a living?"

with "I'm a data scientist" could result in a blank look. Worse, you've lost control of the impression you are making—they'll put you in a box according to their stereotype of a data scientist.

This could be okay if they got the memo that data science is the sexiest job of the 21st century, but it could also mean they assume you meet all their worst stereotypes of a software geek or statistician and then some. Crucially, you won't know which one they landed on.

A better question to answer is the Steve Jobs question "What did you do for [our company] today?" Answering this question rather than the literal question posed by the person you ran into allows you to develop your own narrative of how you help your company.

You aren't even bound by the strict wording of Jobs' question (which was designed specifically to discover the most recent thing a hapless employee had done for Apple)—you are free to answer it based on the most impressive thing you have done over for your company over any timeframe you choose. By reframing the question in your head before answering you've taken control of the opportunity to promote your brand.

To be effective in this arena, you will need to have the answer on the tip of your tongue—to some extent, this means developing an elevator pitch for what you do for your organization.

Although the idea of an elevator pitch is frequently associated with entrepreneurs, most often in pursuit of either sales or finance, the basic concept can easily fit to the task of explaining how you contribute to other workers in your business or potential clients.[8]

The key to a successful elevator pitch is preparation. You need to have a clear idea of what you are trying to achieve, so write down your goal before writing out the first draft of your pitch. Then practice saying it loud to get an idea long it goes for, and to get it to flow nicely. You've got to get in and out within around 30 seconds to succeed, and you've got to have it on the tip of your tongue to work properly.

Remember to record yourself a couple of times and play it back to make sure you've got a decent intonation. Try to avoid speaking in a monotone or speaking too fast, or other problems that mean that you sound like someone reciting a shopping list rather than explaining themselves in a natural voice.[9]

The idea of the elevator pitch is a simple one, and its application doesn't require particular talent, only some effort. However, used in the right place, it can be a highly effective method for explaining how you fit into the big

---

[8]www.mindtools.com/pages/article/elevator-pitch.htm.
[9]www.thebalancecareers.com/elevator-speech-examples-and-writing-tips-2061976.

picture, and how what you do benefits the greater good (while giving the freedom to define the greater good to your own advantage).

At the very least, you will have a better answer when you are asked, "What do you do for a living?" at a party than "I'm a data scientist." More than that, by making you focus on how you make a contribution instead of trying to label yourself, by preparing an elevator pitch you will have a firm foundation for use with any other kind of personal branding exercise you want to consider.

Earning Trust

The highest state you can aspire to in a relationship with the people you are trying to assist is to be trusted. Trust is what ensures that your organization comes to you early with problems that need your help to solve, ensures that they give your solutions a fair hearing and implement them without trying to prove that they won't work first.

Earning the trust of your customers and clients takes the idea of a personal brand further. Although trust is earned repeatedly within each new relationship, it is a deeper level of engagement than having a personal brand and provides a firmer foundation for future work.

Just as for the previous sections, this needs to be done with an eye on promoting people to trust your team and to trust data science as a profession that will aid their company. Trust in either yourself as an individual or trust in the idea of data science by themselves won't be sufficient to ensure they trust your solutions to their problems.

In a later chapter we will see that ensuring that users trust your models is of paramount importance for them to be used. If your users are able to trust you on a personal level, or the model is presented to them by someone they respect who trusts you on a personal level, you will have crossed an important barrier for achieving your user's trust in your model.

At the same time, it is more difficult to get your clients to trust you than it is to get them to trust your model—it's just that if you are successful in getting them to trust you the pay-off is that you need to re-earn that trust every time you want to propose a new model.

Building that trust is something that will take place over a longer period of time, with a different approach. The point here is not that you provide models that do what they say on the tin by attaining a good accuracy measure or even that the models you provide develop a reputation for significantly improving the organization's smooth running.

The famous book on building trust for consultants, *The Trusted Advisor*,[10] defines four ingredients for personal trust—credibility, reliability, intimacy, and self-orientation. A lot of this book has to do with the first two, as these elements can apply to statistical models as well as to humans. The last two really only apply to people.

More generally, the authors of *The Trusted Advisor* are keen to point out that although credibility and reliability have a black and white, technical dimension (which is what I have covered in relation to models in other chapters of this book) they also have an important emotional dimension, which will not map so easily onto models, but is important for your personal credibility when trying to win over your clients or potential users.

Intimacy can be seen as more difficult to achieve than either credibility or reliability. To be credible, you just need to know your stuff. To be seen as reliable, you just need to deliver what you say you will, when you said you would—on the face of it, it's entirely in your control. To establish some degree of personal intimacy with others requires accepting the risk that you will be rejected.

The final ingredient is an inverse relationship to trust—the authors of *The Trusted Advisor* call it "self-orientation," and the less oriented toward yourself your clients perceive you to be, the more successful you will be at gaining your trust.

At least, if you appear to only take your interests to heart, you will find it difficult to win their trust. *The Trusted Advisor* authors could have called this factor "altruism" or "awareness of others" to avoid the inverse relationship.

Although some ways you can be overly self-oriented can be obvious, such as only caring about the paycheck or the kudos of solving the important problem others have already failed at, others are more subtle and pernicious.

Many of the latter could trip up a data scientist—"a desire to be seen to be right" or "a need to look clever." This book was written before smartphones became ubiquitous, so the easiest and quickest way to be seen as paying insufficient attention to your client in 2019 is missing—"phubbing," or looking at your smartphone while talking to others.

It takes time to build up trust, and you can't short circuit the process. *The Trusted Advisor* authors propose a five-step process to gain a client's trust when working on a particular problem.

1. **Engage:** Show the client they have your attention.

2. **Listen:** Show the client you understand their problem.

[10]David H. Maister, Charles H. Green, and Tobert M. Galford, *The Trusted Advisor* (NY: Touchstone, 2000).

3. **Frame:** The root issue is identified.

4. **Envision:** A vision of an alternate reality is sketched out.

5. **Commit:** Steps are agreed upon.

Some of these steps could be seen as overlapping or complementing the design thinking process—envision could be seen as another way of saying Ideate. In some ways, Chapter 1 of this book could be seen as being a guide to engaging and listening.

As mentioned earlier, the element that the treatment in *The Trusted Advisor* emphasizes that is often lost, and likely to be the more difficult for data scientists is the emotional component over the rational component. Some paraphrasing of their advice on listening can illustrate the idea.

The authors identify a number of types of listening, and they also identify the message that the way that you listen sends your client as at least—possibly more—important than the information you get from the client by the process. Listening, then, is an opportunity—yes to learn—but more than that to show by your demeanor and body language that you are on your client's side and that you care that she succeeds.

As mentioned previously, listening can come in a number of types, and the types listed in the book related more to how the listening process affects the person who is being listened to than they affect the listener. For example, "supportive listening," as would be expected, is listening that makes the listened to person feel supportive.

However, essential to all the types of listening that are presented is the need to avoid interrupting the speaker's flow and allowing them to present their story as they seem fit. This may often feel difficult to do, as so often we think that we have heard the story, or one like it, before, and rush to jump in. It takes practice to avoid doing that.

The honest truth of this is that there is a limit to how much of this can be explained in a book. The best way to learn how to do it is to practice and accept that you will fail but also to accept that when you fail it's not the end of the world.

Pushing yourself to push forward with the emotional parts of your arguments, even if it's scary, will enable you to solve more problems with data science, as you will be able to win over more people. By starting small and ensuring your sales pitches target their real needs, you can grow in your clients' trust and become one of their highest valued partners.

Summary

Selling yourself, your model, and data science itself are all fundamental to being a data scientist. Unfortunately, it is not simple to persuade people to use your models, to persuade them that you are to be trusted, or that data science is a valid approach to solving their problems.

Part of the difficulty is that there are often multiple sets of people you need to convince to be allowed to both work on a solution and later implement it. Their needs are likely to be different and what you will need to do to convince them is likely to be different.

The first part of the journey is convincing your clients that your model meets their needs. Building trust in a model is a little less onerous than building trust in a person but some of the elements are the same—a model needs to be credible and reliable. To be credible, the model needs to be understandable, and you need to ensure that your client understands why your model is the solution to their problem.

Selling individual projects is one part of the picture, but marketing yourself and your profession as a data scientist is also important. Data science, for a few reasons, including its relative novelty, the hype surrounding it, and the lack of a specific body representing the profession, is arguably a blank slate from a marketing perspective.

This can be a good thing or a bad thing, as it means that both the responsibility and the opportunity for explaining data science to potential customers and clients is in your hands. Either way, a data scientist needs to re-orient new clients or customers to their own understanding of what a data scientist is, establishing moderate and achievable benefits of data science with their audience.

Your personal brand is another crucial element required for success with clients. This is true both for consultants working with external clients and data scientists working on problems within an organization. There are myriads of techniques suggested for improving personal branding, although some are more applicable to data scientist's situation than others.

An example of a technique with wide applicability is the elevator pitch, and the exercise of focusing on what your personal brand is and distilling to its most important elements to develop an elevator pitch is a great platform for other personal branding activities. For the data scientist, it also helps you ensure that whoever you're talking knows what you mean by data scientist when that phrase means many things to different people.

Being able to achieve a relationship with your client to the point where they trust you will ensure that they not only implement your solutions but also seek your advice early. However, this requires a big commitment, and the determination to not let feeling uncomfortable or the possibility of embarrassment interfere.

The Trust Equation suggested by Maister, Green, and Galford says that human consultants need to be credible, reliable, able to achieve (sufficient) intimacy with their clients, and able to convince their clients they have their best interests at heart. Although the concepts of intimacy and selfishness can't be applied to a model, credibility and reliability can be.

We will explore more specific ways of ensuring that your model is both believable and understandable in order to ensure credibility in the next chapter. In the chapter after that, we will examine ways to ensure your model is reliable by keeping it performing at as close to the way it first performed during testing and validation as possible.

Sales Checklist

- Does your presentation of the model create a link between your model and the solution of the customers' problems?

- Have you considered the emotional aspect of your presentation?

- Do you have an "elevator pitch" answer to the question "What do you do?"

- Have you thought about your own definition of data science, and how it solves your customers' and clients' problems?

- How are you building trust with your clients? What would your clients say if someone asked what kind of person you were?

Believable Models

Earning Trust

It's inevitable, when making a forecast or a model (meant in the broadest possible terms), that you will be wrong—"all models are wrong,"[1] after all. The obvious challenge is to convince your audience that you are doing something they can use, even if your model is wrong.

An essential tool in these circumstances is the ability to explain your model to its audience or users. Without some form of explanation for how the input variables are influencing the model's output, you can't make any kind of hypothesis about what is happening. Without a hypothesis around what the data is telling you, you can't compare the result to existing knowledge. Yes, you will have some sort of accuracy measure, but it will lack the context of how and why it achieved its accuracy.

That, in turn, immediately hampers your ability to use the knowledge of subject matter experts (SMEs) to improve your model as you can't compare your model's view of the problem to the subject matter expert's view of the

---

[1]Box GEP and Draper N, "All models are wrong, but some are useful," *Empirical Model Building*, 1987.

© Robert de Graaf 2019
R. de Graaf, *Managing Your Data Science Projects*,
https://doi.org/10.1007/978-1-4842-4907-9_4

problem. Additionally, the inability to use subject matter experts' knowledge directly in your model is a missed opportunity to win allies for your work and a missed opportunity for improving your model.

At this point I should note that previous chapters haven't assumed any specific prior knowledge, other than some sort of experience building models for prediction. In this chapter, at least some of the discussion assumes a basic knowledge of generalized linear models (GLM) and regression.

I still believe that if you don't have this background you will get a lot of this chapter. However, if the idea that ordinary least squares regression, logistic regression, and Poisson regression can all be seen as examples of the same thing isn't familiar to you, you will get a lot more out of this chapter if you read a basic generalized linear model texts, such as the one by Dobson[2] or Faraway.[3]

To provide credibility for a model, the explanation of what it is doing needs to make sense to the user. If you're presenting the model to users, the best situation is that your model can illustrate at least one relationship they already know about, and another that shows them something new. The first confirms that the model has found relationships which are true. The second shows that the model has found something new that the user didn't know before.

If you can't provide any findings that accord with something that the user has observed themselves, it is unlikely that they will accept your model can be believed. At the same time, if you can't offer them anything new, it is unlikely they will accept that your work is worth paying for.

In fact, it is not enough to simply allow users the ability to understand models. They actually also need the ability to criticize them and ensure that essential expected information is included.[4] We will see more on this topic toward the end of the chapter.

In Chapter 3, I introduced the idea that to build trust as an individual, you need to ensure you have credibility, reliability, intimacy, and minimize self-orientation, as explained in *The Trusted Advisor*.[5] Although the latter two attributes are the domain of humans, the first two can be applied to models. Ensuring your model is both credible and reliable is fundamental to gaining your users' trust. In this chapter, we will concentrate on credibility.

---

[2]Annette Dobson, *An Introduction to Generalized Linear Models* (London: Chapman & Hall, 2002).
[3]Julian Faraway, *Linear Models with R* (London: Chapman & Hall, 2009).
[4]Been Kim, Rhajiv Khanna, and Oluwasanmi Koyeyo, "Examples Are Not Enough!: Learn to Criticise," *Neural Information Processing Systems (NIPS) Conference* (Barcelona, 2016).
[5]David H. Maister, Charles H. Green, and Robert M. Galford, *The Trusted Advisor* (New York: Free Press, 2000).

To be credible, a model will need at least three attributes:

1. **Intelligibility:** Users need to be able to understand the link between inputs and outputs. This link, therefore, needs to be visible, rather than hidden in a black box.

2. **Predictability/consistency:** When your user has seen results or has explored the model definition, they should be able to be in the ballpark of knowing which of two cases is more or less likely to be a particular case, or will be greater if a continuous model.

3. **Reflect knowledge of the real world:** Users will frequently have significant experience of the way what is being modeled behaves in real life.

A fourth point may not be essential but is still useful—the ability to talk about when your model is most likely to be wrong, and to quantify by how much and in what direction it is most likely to be wrong. This is another area, along with the issue of model intelligibility or explainability, that is receiving more attention now than it did in the first rush of machine learning (ML) hype.

If you violate any one of these principles relating to model behavior, your user's trust in your model will rapidly evaporate. Although, especially in the case of the latter two points, there are not always simple ways to ensure that a model fits these rules from the start, there are ways to make it more probable that the model fits these rules, and ways to check if your model fits these rules.

This chapter will discuss each of these attributes in turn. Intelligibility is the most technical of these issues and so needs an especially thorough treatment. However, this doesn't mean that intelligibility is the most important just because of this extra word count needed. My strong opinion is that each of these factors is of equal importance, forming a three-legged stool, where if one of the factors does not exist, the stool will not work.

A lot of this chapter will emphasize nonlinear relationships in a number of ways. The idea that an important way that machine learning algorithms can be more accurate than simple regression models is by representing that nonlinear relationships should be well known to ML practitioners. So too should the idea that a drawback of allowing extra nonlinearities is risking overfitting. Less often discussed is that nonlinear relationships are more difficult for users to understand and are therefore a barrier to trust.

Most machine learning (ML) models don't allow the modeler to select which variables will be modeled as nonlinear effects. An exception is the generalized additive model (GAM), which also allows the nonlinear effects to be visualized in its major implementations. This is obviously more work but it means that nonlinear sections can be confined to the most important variables so that the number of nonlinear variables can be kept to a smaller number that can be explained. We will explore this ability later in the chapter.

As we have seen in earlier chapters, models need to be fit for purpose. Some models will, therefore, need a higher emphasis on how credible users find them than others. Some may need only a very light touch in this area. As discussed in Chapter 3, the point is to consider the users' needs carefully and match the level of validation accordingly. In particular, consider how close the users will get to the results and the cost of misclassification—users will usually be aware of misclassification costs at some level themselves (albeit often only from their perspective) and adjust their tolerance accordingly.

Model Intelligibility

Intuitively, there are two strategies available to build models that are both accurate and can be interpreted. The first is to build a model that is intrinsically both interpretable and work to ensure its accuracy. The other is to build an accurate model and figure out a way to interpret it after the fact.

In the past, it was often assumed that, on the one hand, models which could be interpreted in terms of the way the input variables affected the output couldn't produce an accurate enough result for most people. It was also often assumed that if you were able to produce an accurate enough model, it was almost guaranteed to be impossible to understand how individual inputs affected the outcome.

In the next few sections we will see how both assumptions are being shown to be incorrect, and see different methods that disprove both assumptions.

Models That Are Both Explainable and Accurate

The most obvious way to explain your model is to make it explainable from the start. If it is at all possible use linear regression for a continuous dependent variable, or logistic regression (binomial or multinomial as appropriate) or otherwise appropriate GLM (Poisson, negative binomial, for example) for a categorical dependent variable.

A particular reason that machine learning models have come to be preferred when they may not be necessary is that as people have collected larger data sets, they have come to believe that a larger data set means a more accurate result, without stopping to check whether this is the case. It may be the case with some specific situations, but very often it is not the case.

Using machine learning algorithms as a first resort can sometimes also lead to habits that reduce your ability to make accurate and interpretable models.

As an example, because some algorithms need the data presented to them in a categorical format, feature binning is sometimes recommended as a data preprocessing step in machine learning guides. However, as the binning divisions are arbitrary and discrete, they introduce inaccuracies. Therefore, it is important to remember that building a traditional regression method requires a different mindset than building machine learning. Some lessons that applied to the machine learning approach will need to be unlearned when you turn to a regression approach.

Beyond that, books that teach regression principles don't necessarily teach good modeling practice. In fact, you could say that books that teach good modeling practice in a regression setting are rare. The following principles which have been selected from Harrell give some idea of what is needed to produce a model that is intrinsically both accurate and interpretable:

- The first step to a model, which is both a good predictor and interpretable, is to consider the data that has been assembled very carefully. Do your samples of each input cover each input's range as you will expect to find them when you score the model?

- Carefully interview the subject matter experts. Do you have data which corresponds to the inputs they believe will be most influential?

- Also, consult subject matter experts on variables most likely to exhibit interactions. For example, Harrell compiled a list of likely interactions relating to human biostatistics.[6] Most likely workers in the fields most closely associated with good data science results can write similar lists—for example, credit scoring, insurance, marketing, etc.

- Consider how you will treat missing values. Is the level of missingness low enough that you can avoid treating the data at all, or will you need to impute the values somehow?

- Check the assumptions you made about the distribution of the data. Is your data really a count, so Poisson regression is the most suitable?

---

[6]Frank E. Harrell, Jr., *Regression Modeling Strategies, Second Edition* (New York: Springer, 2015).

■ **Pro Tip** Leveraging your subject matter experts knowledge, where it is available, is an excellent way to create the foundations of a model that is both accurate and interpretable. In Chapter 1 we discussed the right kinds of questions to ask—questions that help identify the most important variables, and the variables most likely to have important interactions are especially useful.

One way to improve the predictive performance of a model is to employ a shrinkage method, such as the lasso or ridge regression. These methods reduce the problem associated with stepwise regression that the variable selection process is discreet, and therefore greedy, which can lead to high variance. Ridge regression, as an example, attempts to reduce this problem by preventing coefficients from becoming too large, hence taking a middle path between discarding variables completely and allowing them to be overly influential.

Linear models can model relationships of greater complexity if the assumption of linear relationships is relaxed. Importantly, the ability to represent nonlinear relationships with little effort is one of the key reasons that neural networks and tree ensembles perform better than linear models, so the ability to relax this assumption is a big step in terms of closing the gap. One of the elements of the strategy for sound predictive models espoused by Harrell is to relax the assumption of linearity for key variables (as determined by knowledge of the subject area).

In essence, by following a careful plan of attack, it is possible to use standard regression or a GLM as required to build a model that is both accurate and interpretable.

The linear assumption can be relaxed by using an additive model to account for influential nonlinear predictors. A generalized additive model (GAM) uses a smooth function such as a spline to represent a nonlinear relationship. The mgcv package in R is one of the most commonly used packages for this model. Its particular advantage is that it provides the ability to visualize the nonlinear areas of the model by plotting the spline relationship. For example, Figure 4-1 shows a plot of a relationship modeled by a generalized additive model.

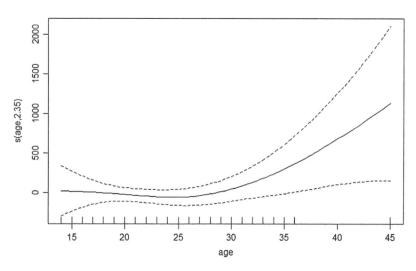

Figure 4-1. A typical plot of a smoothed variable from a generalized additive model prepared by the author from the birthwt data set available from the R package MASS. Note that the y-axis is not the baby's birthweight but, instead, is how much extra weight compared to the typical baby delivered by a 15-year-old mother is due to the mother's age.[7]

One of the biggest advantages of these plots is that they show where turning points can be found. In the case of Figure 4-1, which is taken from a model of babies' birthweight where the x-axis is the mother's age, there appears to be no significant relationship between birthweight and mother's age up until approximately 26 or 27 years of age. From this point on, there appears to be a positive correlation between mother's age and birthweight (although the confidence interval widens as the data thins out, as the number of data points declines with mother's age).

This is a richer view than we would have had with a purely linear model, where the model would have to represent all this information with a single line, by drawing a line with a slope somewhere between the close to zero gradient of the younger mothers' section, and the distinct positive relationship for older mothers. This version is shown in Figure 4-2.

[7]Brian Ripley, Bill Venables, Douglas M. Bates, Kurt Hornick and Albrecht Gebhardt, *MASS: Support Functions and Datasets for Venables and Ripley's Modern Applied Statistics,* 2019, https://CRAN.R-project.org/package=MASS.

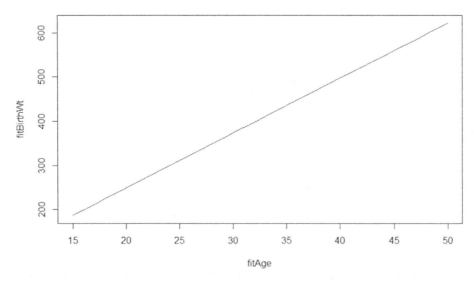

Figure 4-2. A linear representation of the relationship between age and birthweight based on the birthwt dataset in the R MASS package. To match Figure 4-1, the y-axis is the size of the effect of age on the birthweight, not the final birthweight.

As a result, the conclusion drawn can be very different—using the linear-only approach with this data, for example, it is a reasonable conclusion that babies born when their mothers are in their mid-20s can be expected to be heavier than babies born to teenage mothers, but this conclusion is not supported when analyzing the nonlinear output from the GAM. Some texts recommend replacing the curve with a series of straight lines. This can be dangerous as you replace a smooth-curved transition with a sharp transition—you change from a line in one direction to a line in a different direction instantly on a point rather than gradually while curving around a radius. What is more useful is discussing why the turning point is where it is with subject matter experts. I discuss this further in the last section of this chapter.

This is an area that links strongly to the earlier observation that clients need to see something they already know and something they don't know in their model—showing a GAM which confirmed the customer's prior thinking that "Age" was an important factor (something they did know), but enlarged their perspective by showing the effect had a peak or tapered off (something they didn't know) has resulted in excellent customer buy-in for me.

The message here is not to abandon methods such as neural networks or Random Forest, but rather to not default them easily or too early. Even when your project strategy suggests that trading off explainability for accuracy is necessary, be aware that there are a number of ways to see into a black box model. Have them at your fingertips to ensure your customers are engaged by your model and its results—we will look at some of them in the following section.

When Explainable Models Are Unlikely

There are two particular situations where achieving an explainable model is unlikely, with a fair amount of overlap between the two. The first is that the number of variables is too many—at some point over very roughly 20 input variables, the notion of an explainable model fades away as the list becomes too large for humans to reason about. The other situation is where manually developed features don't lead to sufficient performance—computer vision and image identification are the obvious examples of this situation.

It should be observed that in both cases there is a degree of scope to push back toward explainable model—however, deciding not to go that route may still turn out to be the most reasonable course. There are a number of signs that this might be the case:

- There are a large number of possible variables, and none of them is particularly strong.

- When you attempt to visualize nonlinear relationships by creating a GAM or some other way, there are several points of inflection in each variable.

- There is substantial evidence of multicollinearity and the usual remedies don't succeed in improving the accuracy of the model.

Assuming your model meets at least a couple of these criteria, you may have cause to use a black box model such as a neural network or Random Forest in place of an intrinsically interpretable model. Alternatively, you may need to build your model quickly and judge that carefully, as crafting the features needed for a good interpretable model will be too time-consuming.

Windows in the Black Box

Having applied the preceding criteria, you decide that only a black box algorithm will give you the performance you need. The other option is to have an opaque model that performs well but provide another modeling to explain it. An extension to this idea is to make an accurate predictive model with an algorithm like Random Forest that isn't traditionally considered interpretable and use advanced techniques to interpret it.

Surrogate Models

A highly intuitive way to force a black box model to be interpretable is to use its results as the target for a second model using an intrinsically interpretable method, such as decision tree or regression. This approach is called building a surrogate model.

Although the approach has always been available, recently, packages and methods have been built which specifically apply this approach. I will discuss two major types of surrogate models and significant implementations in the sections that follow.

Local Surrogate Models

Using Random Forests to discover quantify relationships as well as make predictions is an active research topic. Recent papers such as "Quantifying Uncertainty in Random Forests"[8] discuss strategies to estimate the size and direction of the effects of particular predictors on the dependent variables over the whole Random Forest, based on U-Statistics.

Drawing on similar themes is the inTrees package in R, which creates a rule set summary of a tree ensemble, and is a great example of one of the several packages that are available in R today to interpret Random Forests, and other ensembles of trees. The inTrees approach is to extract rules from the trees making up the ensemble, and retain the highest quality rules as an explanation or summary of the ensemble, based on attributes such as frequency and error of the rule.

The previously mentioned methods only work with ensembles of trees, including Random Forests and Gradient Boosting Machines. An option which explains results from any algorithm that has emerged recently is the use of universal model explainers, of which Local Interpretable Model-Agnostic Explanation (LIME) is potentially the most prominent example.

In contrast to a method like generalized additive models, LIME will provide an explanation on a case-by-case basis. That is, for a set of parameters representing a case to be scored, the LIME explanation represents how the different variables impacted on that particular case; if another case is presented to the algorithm, the effect of variables can be quite different.

The explanations are presented as a horizontal bar chart, showing the relative size of the influence of different variables, with bars extending to the right for variables that make the classification more likely, and to the left for variables making the outcome less likely. At a high level, the effect of the variables comes from a sensitivity analysis, which examines the classification results of other cases closely similar to the case of interest.

This is the local aspect of LIME—explanations are given on a case-by-case basis, rather than providing rules or a guide to the model as a whole. This is a notable point of difference to the methods discussed earlier for ensembles of trees. Additionally, LIME is currently only suitable for classifiers, rather than regression models.

---

[8]Hooker G and Mentch L, "Quantifying Uncertainty in Random Forests via Confidence Intervals and Hypothesis Tests," *Journal of Machine Learning Research*, 17, 2016.

Global Surrogate

LIME is an example of a local surrogate—effectively a linear model is built that works in an infinitesimal area around the selected example. The intuitive opposite of a local surrogate is a global surrogate—a surrogate model that is expected to explain how the underlying model behaves across the whole of its domain, rather than at specific points.

Local surrogates and global surrogates have the potential to draw quite different relationships between the input and output variable if there is significant nonlinearity or interactions effects. Therefore, global surrogates may be especially useful when attempting to explain the overall operation of a model to an audience of subject matter experts, as we will discuss in more detail in the following text.

The difficulty with the global surrogate is that you effectively need to build an extra model, with all the pitfalls you encountered building the first one. The surrogate model may itself introduce distortions in the way you understand the underlying model, on top of distortions that were introduced by creating the model, to begin with.

Decisions trees are arguably the most popular algorithms used to build surrogate models. Although in theory it should be straightforward to build a decision tree model to fit outputs, in practice it may still be messy. Christopher Molnar has created[9] a function, TreeSurrogate, as part of his iml ("interpretable machine learning") package in R, which simplifies the process of fitting a decision tree from the PartyKit package to your predictions.[10]

Whether you decide to use the TreeSurrogate function or not, growing a single decision tree model with the results of a black box model as the target should be one of the arrows in your quiver to explain a model that initially appeared uninterpretable. No doubt, with the growing interest in ensuring that models can be interpreted, more options will begin to appear in the near future.

The Last Mile of Model Intelligibility: Model Presentation

Creating a model that an end user can understand means, on the one hand, ensuring that they understand at a basic level what the input and output variables in the model are, and on the other, that they understand how those variables operate within the model.

---

[9]https://cran.r-project.org/web/packages/iml/vignettes/intro.html.
[10]https://cran.r-project.org/web/packages/iml/iml.pdf.

In each of these cases the final presentation is crucial to ensuring the end goal of seamless user experience is met. While it is relatively obvious that input variables that are themselves the output of complex models or variables with opaque names such as "Var1" will add to user confusion rather than aiding their comprehension, sometimes it isn't obvious how much explanation is needed.

Part of the problem is that from the modeler's perspective what's important is what goes into the variable—the name of a ratio input to a model might well refer to the variables that make up the ratio. Obviously, this won't mean that much to the users of the variable. Think of the names given to accounting ratios—the "quick ratio," the "acid test," or to dimensionless constants in physics and engineering, such as Reynolds number (which, at a simple level, expresses the degree of turbulence in a fluid).

Reynolds number, while not possessing a perfectly communicative name (from that perspective, "Turbulence number" might be an improvement), does illustrate the notion of divorcing the calculation of a model input from its meaning within the model in a different way—there are multiple Reynolds numbers for use in different engineering models in different contexts, but essentially using the Reynolds number in the same way—to quantify the turbulence of a fluid. There is a Reynolds number for flow in pipes and channels, for a particle falling through a fluid, for impellers in stirred tanks, and for fluid flowing in a packed bed—all with different calculations, all expressing, at least at a basic level, the same concept.

In data science models this plays out in two ways. One is that if you want to reuse your model in as many places as possible it's handy to label your variables in a way that is independent of their ingredients for portability. In a credit risk model, for example, income or assets could have differing calculations for local taxation or other regulatory reasons. Therefore, you would either need to identify the regulation your definition complies to, or provide enough detail that your user can do their own check.

The second is more to the point—your interface will require labeling if others are to use the model, and labeling that refers to the variable's meaning within the model does more to explain the input to users than labeling explaining its ingredients.

In the case of Reynolds number the explanation "a constant expressing tendency towards turbulence, where a higher number tends more towards turbulence" is more useful, and does more to explain its purpose in the models it is found in than "the ratio of diameter, velocity and fluid density to fluid viscosity."

For your models, explaining an engineered variable in a way that users can understand it will seldom simply mean listing the underlying variables or sticking the piece of algebra that is its mathematical definition onto a web page. It means explaining the physical meaning of the new variable in the context of your model, and if you don't know what the variable means in the context of your model it's time to talk to your subject matter experts (and if they don't know, you may need to drop that variable).

The bottom line is that you cannot expect your users to be mind readers, and not only does strict attention paid to labeling lead to users being more likely to use your model, but it also means they will be less likely to make incorrect inferences from your model's results.

Standards and Style Guides

There are a number of resources that can be used as guides on how to develop standardized systems for labeling variables. Some of the most popular include Hadley Wickham's R Style Guide and the Google R style guide. However, many of these focus more on the capitalization of words or when to use curly braces vs. square braces than on the usability of variable names.

Another place to look for advice on the best way to name your variables is from Clean Code practitioners. Advice such as "avoiding disinformation in names" is a good start. A particular piece of advice from Robert C. Martin, paraphrased, is that the more often a variable appears, the shorter it should be. Conversely, variables that appear infrequently require longer names that virtually define them.

This advice is still specific to programming, however. We can possibly find other advice that fills in more of the gaps.

Better is advice from the database and data warehousing experts, who are more strongly concerned with the task of ensuring that nonexperts are able to understand the data that they are presenting.

- Each variable should have a unique name within the database (or model).[11]

- The name should have a maximum of 30 characters if not limited further by the modeling environment.

- Use one abbreviation throughout all variables if needed. For example, if you use "weight" in more than one variable, consistently shorten to "wt."

- Use single case for nouns and present tense for verbs.[12]

---

[11]"SQL Server Naming Guide and Conventions," accessed April 4, 2019, at www.cms.gov/research-statistics-data-and-systems/cms-information-technology/dbadmin/downloads/sqlserverstandardsandguildelines.pdf.
[12]"Data Warehouse Naming Standards," accessed April 4, 2019, at https://kb.iu.edu/d/bctf.

Keeping to an agreed set of rules within your team will ensure that users are able to fully understand, both the way that the variables affect the outcome and what those variables mean.

Model Consistency

The second attribute of a model that users find credible is that it is consistent. Most users' default position is that if getting older increases the risk of death at one stage of life, it ought to increase the risk of death in other stages of life—if getting older by a year suddenly makes you less likely to die, they are left scratching their heads.

Models that take into account interactions and nonlinear effects are likely to provide more accuracy. They are also more likely to be subject to effects that are artifacts of the data, rather than represent the ground truth of the scenario being modeled. This is the familiar story of overfitting.

Overfitting in aggregate is well covered in machine learning texts, with methods such as cross-validation being recommended to detect it, and methods such as regularization being recommended to prevent it. However, these methods are mostly intended to pick up models that overfit on average. They are less good at picking up areas in the model that are different from the overall picture, and therefore likely to undermine a user's confidence in what the model is saying.

This is an area where it is important to proceed carefully. If you selected a model that allowed nonlinear relationships, you must be working under the assumption that some of the time those relationships accurately represent what you are analyzing. The problem is twofold—on the one hand, there is a risk that some of the nonlinearities aren't real. On the other hand, you will often find that even if the nonlinear relationship is real, your users will have difficulty accepting it.

A particular problem is relationships that change direction, as highlighted in the opening example. The technical name for relationships that maintain the same direction is monotonic—the opposite is nonmonotonic. Nonlinear models are not necessarily nonmonotonic, and linear models can also be nonmonotonic (think of square terms) but they are more likely to be monotonic, and you have less control over intrinsically nonlinear models compared to adding square terms or similar to a linear model.

Users are likely to expect that relationships hold over the whole range of the model. Hence, even if there is a good reason for a relationship to run in different directions at different points, it may still be an uphill battle convincing your users.

Work through to the explanations. If you say to someone, "getting one year older makes you more or less at risk of death in the following year?" they will almost always reply "More likely." If you tell them there's a war in progress, and compulsory frontline service for every aged between 18 and 30, and ask them whether a 29-year-old is more likely to die than 31-year-old under those conditions, they will answer "the 29-year-old."

The point is that people will believe the nonlinear relationship if it makes sense and they understand it. The task for modelers is to verify with subject matter experts whether or not there is a real cause for the nonlinear behavior, and then to communicate it to the users.

Communicating the cause to users can occur in a range of different ways. We will see some of them in the chapter to follow on communication. For now, it is important to realize that you should try to keep the number of novel nonlinear relationships—those where you are teaching your users about a behavior for the first time—to a manageable number.

In the next section, we will look at workshops to engage subject matter experts, and get their opinion on whether nonlinear relationships are real or spurious.

Tailoring Models to the Story

Many models are subject to the flat maximum effect—the optimum isn't a point, it's a region of alternative ways to achieve the same value. Thus, a model reaches a point of optimization in its development where it becomes difficult to make any further improvement on accuracy, even with significant changes to the inputs.[13] This can be viewed in a pessimistic way as the data science version of the law of diminishing returns found in economics, and when first proposed, this was exactly how it was viewed.

However, there is a glass-half-full version of this idea. Rather than focusing on the flat maximum as a limitation on model performance, you can instead focus on the flat maximum as providing multiple pathways to the best possible result. Therefore, a data scientist is in a better position than, for example, an engineer who often needs to trade in performance for cost or performance against metric A for performance against metric B.

Instead, once a data scientist has optimized a model against accuracy, they have the freedom to look for alternative models that optimize other metrics. Examples in the literature show models which are more complicated and more specific becoming simplified and more general.

---

[13]David Hand, "Classifier Technology and the Illusion of Progress," *Statistical Science,* Vol. 20, No. 1, 2006, https://arxiv.org/pdf/math/0606441.pdf.

However, in the context of this chapter, the flat maximum effect provides a great opportunity to choose a model which matches the user's view of the world most closely. You will realize that the model is going to be accepted by users far more readily than any others. The other opportunity the flat maximum gives us to make a model that matches your customer's target market, for example, a risk model that determines that too large a proportion of a credit providers target demographic is an unacceptable risk will hinder rather than help their business, and a model that is more selective within the intended demographic but is just as accurate could be better received.

The opportunity the flat maximum effect provides is the opportunity to challenge the idea that there is one true model that can be found, which often has the side effect that once the target is known, models are created without considering how a typical user understands the relationship being modeled—if there is a "true" or a "best" model, considering the users' view is unnecessary.

In reality, as there are multiple optimal models, there is room, obviously subject to the correct data being available, to build a model that takes the users' viewpoint into account, confirming and codifying it wherever possible.

Meet Your Users to Meet Their Expectations

Models are meant to be used, and users need to both believe them and understand them to use them to their best advantage—and may not use them at all if they don't trust them. One way to gain your users' trust—and tease out important subject matter expertise to better your own understanding—is to present the model in person and sanity check the relationships.

To get the best out of your subject matter experts, if possible, the last step of your modeling process should be to invite their feedback on both the output of the model and the way the input variables work together to achieve the result.

The LIME visualization's format is a useful guide to how to visualize the effect of the input data on the output. As LIME visualizes the partial derivatives of all the input variables at a point represented by a particular case (a particular borrower, a particular insured, a particular consumer, etc.) within the scope of your model, this visualization gives you the opportunity to ask questions in a way that make sense to the subject matter expert. Visualize representative cases to show your users how the model works, and ask their opinion of whether the model makes sense to both get their buy-in and to validate the model.

Explaining How Models Perform

One of the advantages of traditional linear models over machine learning models is that a wide variety of tools have been developed to assess their performance.

A particular example is that the confidence intervals of both parameters and predictions of linear models are readily and easily available from almost all statistical packages, and the calculation method is relatively simple and widely covered in standard textbooks.

This is far less the case in the case of machine learning algorithms, including decision trees, Random Forests and other ensembles of trees, and neural networks. In the case of these algorithms, texts and packages are far more likely to discuss outputs as if they were purely deterministic, without error bars or similar. In statistical jargon, models return only a point estimate—of the target value itself, of a regression problem or of the classification probability, of a nominal Boolean target.

There is, however, no reason not to use confidence intervals or prediction intervals to describe the results of machine learning models, and both a growing recognition of the need to do so, and a growing range of tools to allow a practitioner to do it.

Similar to the situation that exists with model interpretation methods, there are methods to calculate confidence intervals for specific kinds of algorithms, and there are universal methods to calculate confidence intervals. I will take a high-level look at a couple of these methods, giving a flavor for how they work and their limitations, rather than attempting to provide an in-depth explanation for another relatively technical topic.

Representative Submodels

In the specific case of Random Forests and Gradient Boosting Machines, both of which are ensembles of decision trees, an intuitive way to construct confidence intervals is to analyze the decision trees making up the overall model.

As these models are intrinsically built up of other models, an intuitive method of estimating the error on the prediction is to derive it from the range of predictions that come from the underlying models.

For this class of models, out of bag sampling is frequently used to quantify the prediction error of the model. The out of bag estimate is derived from trees that didn't use a specific observation, with the error being computed for each case. The intrinsic availability of the bagged trees, therefore, makes it simpler in the case of Random Forest and its cousins to compute prediction intervals.

As a result, many implementations of Random Forest have the out of the bag estimate of prediction error readily available as an out of the box part of the package.

Variable importance is also readily available for Random Forests and tree-based cousins such as Gradient Boosting Machines, by totaling the importance of each split in each tree making up the ensemble of trees. Variable importance is often used by practitioners to identify which variables to leave in or out, but a better use of them is in discussion with subject matter experts to decide whether the model is using the correct logic.

As mentioned previously, this is an area of active research, with new implementations of ideas frequently appearing in R and Python as they emerge—the paper "Quantifying Uncertainty in Random Forests" is but one example.[14]

Model Agnostic Assessment

A far more general way to approach the task of creating confidence intervals for machine learning models is to use bootstrap confidence intervals.

The bootstrap is essentially resampling with replacement. To bootstrap your model's output, you need to create multiple models that are slightly different. Intuitively, the simple way to that is to create multiple models, each using a different random sample of the original data set. Once you have created a sufficient number of models you can then compute the error for each one, and develop a confidence interval.

Being able to discuss the limitations of your model is an important way to ensure that the model users can trust your results. Prediction intervals themselves are a relatively statistical method of quantifying your model's limitations. You will need to consider your audience carefully when deciding how to communicate what the confidence intervals are saying.

Both cross-validation and bootstrapping can be used to derive prediction intervals for any form of the model. Cross-validation tends to overestimate the prediction error whereas the bootstrap tends to underestimate it.

More recently, there have been developments in using variational calculus as another alternative method to compute prediction intervals for machine learning models. Although the details don't fit here, this illustrates the point that this is an area attracting increasing interest, and is another example of a method of computing these prediction intervals.

---

[14]Lucas Mentch and Giles Hooker, "Quantifying Uncertainty in Random Forests via Confidence Intervals and Hypothesis Tests," *Journal of Machine Learning Research*, 17, 2016, p. 1–41.

In some ways it is more informative for the user to know how the prediction interval varies across the model's output rather than knowing an absolute value. Knowing that the prediction is narrowest (a proxy for when the model is at its most confident) at the center of the range of predictions the user is interested in gives the user maximum confidence in the total output of the model.

This may make residual analysis of your model's output as important as the overall error analysis. Although residual analysis is commonly used in the context of GLMs to validate assumptions such as constant variance (which, if violated, may indicate an incorrect distributional assumption) if the residuals fan out at the ends of the range of the predicted variables or for the range of a particular input variable in a machine learning context, it could indicate that in that region the model isn't sufficiently accurate, potentially owing to a lack of data in that region.

You can also perform a "soft" kind of residual analysis by comparing the prediction error for particular regions of your data, for example, male vs. female, smoker vs. nonsmoker or child vs. adult, if we imagine medical-related model.

Taken together, being able to quantify the certainty of your model in a way that is meaningful to your users, and being able to explain when your model works at its best are key to being able to maintain your user's trust. A clear view of your model's limitations, rather than putting people off, helps them to know when they can use it most effectively and with most confidence.

A Workshop with Subject Matter Experts to Validate Your Model

The final of the three attributes required of models that I listed at the beginning of the chapter was that models need to match your users' experience of the real world. If users believe that age or gender have a particular effect on an outcome, either the model needs to align with that position, the modeler needs to be able to state that the effect is too weak to be seen within the model, or finally the modeler needs to have enough evidence to get people to change their minds.

Research has shown that being involved in the solution to a problem dramatically increases the probability that people accept the solution. Hence, even if you are coming into the situation with your own favored solution, presenting the problems that this solution is expected to overcome—without mentioning your solution—will help ensure people come around to your view. Even better if your users can be active participants in the conversation rather than people you talk (down?) to, the latter scenario allowing them to cement their own negative views.

While obviously this depends on having built a white box model that can be discussed in this way, the rewards are great. For example, it is the best way to work out the final variable names that the users will see because the users can propose their own meaningful names. It allows you to avoid the "Google Flu trends"[15] mistake of presenting a model that includes obviously spurious relationships. Finally, the mere fact of asking your users' opinions ensures their buy-in, and listening to their advice and making changes accordingly will seal the deal.

At the same time, as has been mentioned earlier, not engaging with subject matter experts, whether they are part of your targeted user group or not, represents a massive lost opportunity from the perspective of getting your model fully validated.

As subject matter experts think in concrete terms, to facilitate a workshop that allows them to have their say requires that you provide concrete examples of the model's output with cases that will resonate with them.

The LIME visualization style is a good way to visualize cases for this specific purpose because as mentioned earlier it displays the local effects of movement in each variable. This allows you to frame your questions for the subject matter experts in a way that means they have no need of any statistics knowledge to answer. Here are some possible examples:

- Looking at this case, would a change from male to female make a positive or negative difference or no difference?

- Would you expect case A or case B to be more likely to lead to a positive result (for a classification problem)/ greater value (for regression)?

- Do you expect that the variables work the same way across the whole domain of the model? (To put it another way—are there expected interactions, such as between gender and the effects of drug dosage?)

Although examining the localized interpretations gives a clear idea of how real-life cases work, they can lead to not seeing the wood for the trees. In a vanilla linear model with none or few interactions, this isn't a problem as there will be little or no difference between the way the model operates in different local regions.

However, where there are significant nonlinear and interaction effects, there can be severe differences. If this is the case, the workshop needs to include

---

[15]David Lazer and Ryan Kennedy, "What We Can Learn from the Epic Google Flu Trends Failure," *Wired*, October 1, 2015, www.wired.com/2015/10/can-learn-epic-failure-google-flu-trends/.

visualizations of the global relationships to account for different slopes in different areas of the prediction space. For important variables, the GAM plots discussed may be a useful way of visualizing the way that key variables change gradient over their range.

In my experience, when people are given these plots, they are often surprised by how nonlinear the relationships really are—people are wired to have a strong expectation that relationships are linear. Giving an influential group of users the chance to look at these plots is, for this reason, a crucial method for ensuring that users buy into your model properly.

Importantly, their responses to those charts will enable you to understand your model better and to draw useful boundaries around its field of operation.

For example, in a session like this I conducted, the x-axis was loosely speaking age of an asset. There was a noticeable turning point when age was plotted against the output variable, leading one of the subject matter experts to realize that it correlated with a change in regulations that affected the way that assets were maintained.

That is, when the asset reached a particular birthday, regulations demanded extra maintenance take place so that paradoxically the asset became less prone to problems at this point. This eventually meant that the scope of the model was changed so that the regulatory change no longer had an effect on the model.

After this session, in addition to finding a concrete action that improved the model, subject matter experts who had the respect of the wider group of users had a better understanding of the way that the model operated. This shows how engaging with users can lead both to a model which is genuinely better, and also held in better esteem by the people who are expected to use it.

Even where you don't have the ability to visualize a relationship to the same degree as you can with a GAM, even knowing the importance of different variables within the model gives you the information you can use to generate discussion and challenge your subject matter experts. Few implementations don't provide variable importance plotting, and there are packages that can plot variable importance for any given model.

Sometimes just knowing that the model is heavily relying on a particular variable that the SMEs consider to be less important is enough to detect a problem.

This has now been demonstrated in a research setting, with Caruana[16] et al., who created GAMs and GA²Ms (GAMS with two-way interactions) for

---

[16]Rich Caruana, Yin Lou, Johannes Gehrke, Paul Koch, Marc Sturm, and Noemie Elhadad, "Intelligible Models for Healthcare: Predicting Pneumonia Risk and Hospital 30-Day Readmission," *Proceedings of the 21st ACM SIGKDD International Conference on Knowledge Discovery and Data Mining*, 2015.

predicting pneumonia case readmission. The focus of the authors was on producing models which are "repairable"—that is the model's reasoning could be inspected by a Subject Matter Expert, challenged, and where required, repaired.

In many cases, the subject matter experts were able to trace unexpected rules to specific cases or clusters of cases, and determine that the rule was a data artifact, not a true medical reason changing readmission risk. The models made with GAMs or GA^{2}Ms were deemed repairable as a rule that was found to be suspect could be removed from the model without knock-on effects to model bias.

The eyes of subject matter experts can detect inconsistencies that are impossible for a data scientist. Although data scientists are expected to understand the context they work in thoroughly, they can't think like the subject matter experts in that area, and often won't notice issues that are obvious to a genuine subject matter expert.

Hence, for example, although a statistician who specializes in medical research will have a thorough knowledge of their area, they will never think like either a doctor or a full-time medical researcher, and a doctor or medical researcher will always notice something different to the statistician when reviewing research results. This is true across the range of applications currently targeted by predictive modeling.

External Validation

In all walks of life, one of the best ways to ensure that the kind of mistakes that become invisible when we have been working on something deeply for a long time are detected and removed is to find a fresh pair of eyes. This is as true of building predictive models as it is of designing a bridge or writing a book.

An example of a specific problem that can be detected by another set of eyes with a statistical mindset is data leaks. It is very easy to create a model with exaggerated performance due to training on data that isn't available to the model at scoring time. This is colloquially known as a data leak, and at least some of the time, data mining competitions such as those on Kaggle are won by people who exploit these leaks.

If a data leak has occurred during your training process, without you knowing it, you will likely have a model that performs very well during your validation, and very poorly when put into service. In this situation, a fresh pair of eyes can save you a lot of embarrassment.

If your organization is large enough, it may be that there are enough data scientists in a team that they can be divided into smaller groups that can work in parallel without seeing a great deal of each other's projects while they are being worked on. In that case, you can get those subteams to validate each other's work, assured that they will have a fresh pair of eyes.

This might be a relatively rare scenario, however. If you aren't in this position, for models that have a great deal of exposure, you may want to consider retaining an external statistical consultancy for validation. This may be particularly important if your model is directly exposed to customers, or if the results are directly incorporated into the way your organization makes a profit.

Both of the second two factors can be checked through this method. Simply by asking you questions and trying to understand what the model is doing, a competent external reviewer will highlight any areas where either the model is a poor reflection of the way your context operates in the real world, or where the model isn't consistent with itself.

This last validation, especially when coupled with a workshop involving subject matter experts, is the most effective way to ensure your model is consistent with itself, interpretable by its users, and makes sense within its problem domain.

Summary

We have seen that as machine learning takes on more complex challenges, where there is a greater cost of being wrong, users have begun to demand greater assurance. To gain that assurance, users now expect to gain a greater understanding of the way a model treats the inputs to achieve the outputs.

The most intelligible models are created via linear regression. Traditionally, machine learning practitioners have assumed that simple regression models or even generalized linear models don't achieve sufficient accuracy, and we reviewed modern techniques which may mean that better accuracy is achievable. In particular, techniques such as generalized additive models mean we can relax the assumption of linearity mean that less well-behaved data sets can be used with regression and still lead to an accurate model.

Regression models are still limited by their assumptions; messy data sets with weak signals may still need a black box machine learning algorithm to achieve the needed accuracy. For these cases, there are now methods of interrogating the black box and developing an intelligible model.

Nonlinear and, especially, nonmonotonic relationships may appear counterintuitive to the user and require additional analysis and explanation. The generalized additive model also provides a good avenue for visualizing such nonlinear relationships to aid the explanation.

Simply making the model intelligible in terms of the relationship between input and output will not always be by itself enough to ensure users trust the models' results.

One area requiring attention is whether the input and variables themselves are intelligible. Naming conventions are important in this regard, as is ensuring that proper definitions are easily available.

You can help users understand and trust a model by allowing them to play with the model themselves, interrogating it to decide whether its answers make sense. This can be achieved through computer simulations that run the scoring engine with user-selected values or it can be achieved through a detailed analysis of the rules that make up the model, where this is available. Being able to obtain an endorsement of this kind from expert users is becoming increasingly recognized as an important way to ensure model integrity and reliability.

It is also important to involve the model's future users, not just to improve the model but also to get the buy-in of influential future users. One way to do that is to hold workshops with your local subject matter experts checking that the inferences from your model match their understanding of the subject—you won't get them to believe your model unless either your model matches their expectations or your data and analysis actually changes their expectations.

Another important way to check your model makes sense is to ensure it is validated by someone who wasn't involved in building it. In a smaller company, this may mean that you need to retain an external modeling consultant.

In Chapter 5, we will look at how to ensure that your models continue to make good on their initial promise after implementation and beyond, including model maintenance and surveillance.

Model Credibility Checklist

Interpretable Models

- Given the type of data you have, and the degree to which it is likely to violate typical assumptions, could you develop a sufficiently accurate model that is intrinsically explainable, such as a GLM or decision tree?

- If not, is there an out of the box tool for interpreting the results of the black box algorithm you are using?

- Can you give your users access to explanations, both at the local level and at the global level?

- Can you explain the level of uncertainty in your model's predictions to your users?

Model Presentation

- Do the variables in your models have meaningful names?

- Can your users easily find an explanation of any ratios or other formulas used for derived variables?

- If your users access the model through a web-based platform or an app, does the UI/UX design allow users to quickly access definitions and explanations for the target and input variables?

Models Consistent with Themselves and Their Subject Matter

- Have subject matter experts had an opportunity to review your findings, both on whether the individual results make sense, and whether how they relate to inputs makes sense?

- Have you reviewed the important nonlinear relationships with subject matter experts to ensure they make sense within what is known about the topic?

- Have you shown the model to another group of data scientists to review whether the model's results are robust and repeatable in statistical terms?

- Have you compared the subject matter experts' opinion of what the most important variables are with the variable importance found within the model?

- Have you removed or repaired any relationships that appear flawed after being reviewed?

Reliable Models
Maintaining Performance

In Chapter 3, we first encountered the trust equation, whose parameters include credibility, reliability, intimacy, and self-orientation. Although you can't be intimate with a model, and no one expects a model to think of other people, a model can be both credible and reliable or it can fail to be either or both of these things. In Chapter 4, we explored how to make a model believable, but it is also important that a model is reliable.

The notion of a model needing to be reliable has received more attention in recent times, at least in part, thanks to the fallout from the global financial crisis (GFC), where models gave poor results for a number of reasons including being used outside the parameters intended by the model designers, and being assumed to continue working even when the input data changed.

The aftermath of the global financial crisis (GFC) ushered in a newly skeptical attitude toward the use of models, some of it warranted, and some of it not. Most recently, partly as a reaction to data science and Big Data hype, authors from within data science such as Cathy O'Neil[1] have begun to warn people of the effects of negligent models that harm people as a side effect or even a direct result of doing their job. Those warnings have also begun to be picked up and repeated in articles meant for a general audience.[2]

---

[1]Cathy O'Neil, *Weapons of Math Destruction* (New York: Crown Press, 2016).
[2]Dave Gershgorn, "Tech companies just woke up to a big problem with their AI," *Quartz*, June 30, 2018, qz.com/1316050/tech-companies-just-woke-up-to-a-big-problem-with-their-ai/.

© Robert de Graaf 2019
R. de Graaf, *Managing Your Data Science Projects*,
https://doi.org/10.1007/978-1-4842-4907-9_5

One of O'Neil's points is that models that don't receive feedback can go awry in a variety of ways. Moreover, in many cases, in the examples she uses there were frequently multiple opportunities to set models that went awry back on the right path or make the decision to discontinue the model. Viewed from that perspective, these examples are highly illustrative of the need to watch and maintain models to ensure reliability.

Unfortunately, users often expect that models should be "set and forget," and don't instantly understand the need to maintain and retrain models. It is your role as a data scientist to set their expectation to the right level as early in the process as possible.

Maintaining users' trust in models with that as a background can be difficult. In a way, it should be difficult—sometimes users who are too eager to believe the good news story coming from the modeler can make it difficult for the modeler to present an appropriately skeptical view of their model's performance.

A lot of the foregoing relates to the problem of maintaining credibility, while also dealing with users who are too quick to assume your models are credible. The next stage of the trust equation, reliability, also comes into play, as models shouldn't automatically be relied upon to provide results at the same level of accuracy indefinitely.

Taken together, these factors imply that although some effort is required to ensure that models run to the optimum performance over a long period, it will sometimes be difficult to persuade users to allow you to keep them updated. That is absolutely my experience, and other authors have also noted that data science as a discipline lags behind the norms of other disciplines when it comes to activities such as quality assurance,[3] which are considered essential in other contexts for ensuring users and customers get what they expect.

Therefore, in this chapter, along with discussing how to keep models in peak condition, I will also discuss how you can persuade your users that this is needed, without making them believe that your model was in some way deficient, to begin with.

At the end of this road, lies the question, when should a model be retired? The answer to this question harks back to our very first chapters where we discussed how to determine the problem you should be solving. The risk is always there that the point of your model may no longer exist because the problem it is meant to solve is no longer relevant.

---

[3]Irv Lustig, "Bringing QA to Data Science," *Software Testing News,* October 11, 2018, www.softwaretestingnews.co.uk/bringing-qa-to-data-science-2/.

What Is Reliability

It's hard to talk about making your models more reliable without deciding what reliability is. Before I move on to a technical definition for models, it might be useful to first think about how it works out for humans.

In the context of the "Trust Equation," reliability is effectively defined as an advisor who is consistent and dependable. Therefore, they are someone who does what they say they will do, what is expected of them, and doesn't act in unexpected negative ways.

These attributes can be carried over to predictive models with a little adaptation. A model or data product should also do what is expected of it, do what it does consistently, and not behave in an unexpectedly negative manner—effectively, it should not have negative side effects.

Therefore, a model can be considered reliable if it continues to perform at the same level as it did when first developed, and continues to make the same decisions given it is presented with what its users consider to be the same data—although this may not necessarily be identical data from the strict perspective of the model training set.

In this context, we consider the model itself (i.e., the set of rules produced by some sort machine learning algorithm), the data that feeds it, and their collective implementation as a complete system which together make up the model.

Although the model or rules are stable and are likely to perform as expected on cases presented during the development phase (although other cases might present a problem), the data is subject to change. Moreover, the implementation though not subject to change in the same sense may provide opportunities for errors to creep into the system. Therefore, generally speaking, these aspects are more likely to be areas which cause failures in reliability.

Data quality has become a particularly fertile and also fashionable topic of discussion as the use (and possibly abuse) of data in commercial settings has expanded. Certain criteria have been suggested as making up the components of reliability[4]—I suggest that they more generally make up the concept of reliability as it applies to models.

1. Accuracy

2. Integrity

3. Consistency

---

[4]Li Cai and Yangyong Zhu, "The Challenges of Data Quality and Data Quality Assessment in the Big Data Era," *Data Science Journal*, 14, p. 2, May 22, 2015, DOI: https://doi.org/10.5334/dsj-2015-002.

4. Completeness

5. Auditability

The criteria listed, from the article by Cai and Zhu,[5] also presents the idea of relevance. I suggest that relevance can be folded into reliability for our purposes, as model users make an implicit assumption that model results are relevant to their situation—if they turn out not to be, they will experience the result as a loss of reliability.

Bad Things Happen When You Don't Check Up on Your Models

Some of the worst outcomes can occur when models are implemented, then left without considering whether what was true yesterday was still true today.

Some of the most notorious examples of this phenomenon have their origin in the GFC. During a crash, previously uncorrelated variables become correlated, as a piece of news on the way down has a negative interpretation. The tendency of stock prices to become correlated during extreme effects was, in fact, a known problem, however, it had been neglected during the period when returns were more benign.

In fact, during the GFC, and other extreme events, multiple models of behavior ceased to work correctly, as the relationship between the dependent variable and the independent variable was disrupted. Probability of default models for both retail and corporate credit is an additional important example of this occurrence.

As a consequence, it is common for models that performed well in standard conditions to become very poor during a financial crash. If the team modeling the events had not developed the model under crash conditions ahead of time, then they would have had great difficulty in developing a new model that functioned adequately in time to be useful.

The lesson here, moving beyond the world of finance, is that models working correctly are heavily dependent on their environment remaining sufficiently similar to the environment under which they were developed. To ensure that models continue to have the correct environment to work under means carefully observing the environment so that changes that would mean that new models are required can be detected early enough so that a faulty model is not used.

It was said soon after the events of the GFC that the events were "black swans," which is by way of saying they were difficult to predict within the

---

[5]Ibid.

framework of existing models because they were outside the modelers' experience or data sets. That is something of a cop-out—research available prior to the GFC already showed that the asset correlations during previous crashes changed substantially, so at least the idea that the models were likely to fail during a crash was known among researchers and available to anyone who was interested enough to find out.

Like unhappy families in Tolstoy, models all have problems in different ways. Some of the problems may be small; others large. Sometimes the problems can actually mean the model is doing harm somewhere. Other times the problems simply mean that the performance slowly degrades until the model is no better than a dart board.

What these problems do have in common is that if you want to find them before it's too late, you need to make a conscious effort to find them, and you need a structure and a process to ensure that you find them.

The flow on effects of whether you get this right or not can be large. The idea that models can't be trusted has begun to have a life of its own, and even while articles promoting the benefits of data science continue to come out, articles warning of the dangers of data science going wrong proliferate.

In Chapter 3, we discussed the need for data scientists to be brand ambassadors for the profession of data science as a whole. One of the most practical ways to do this is to create models that both maintain their performance over their lives and can be trusted to both fulfill their original mission, and to do that without causing collateral damage.

Benchmarking Inputs

A model's performance is determined by the quality of the data it's fed. Data quality can have many dimensions,[6] which to some extent can be determined by the way the data will be used. Some authors have identified their own lists of more than 20 possible dimensions. However, six dimensions are in particularly common use:[7]

- Completeness

- Uniqueness

- Timeliness

---

[6]Fatimah Sidi, et al., "Data Quality: A Survey of Data Quality Dimensions," 2012, *International Conference on Information Retrieval and Knowledge Management,* DOI: 10.1109/InfRKM.2012.6204995.
[7]*The Six Primary Dimensions For Data Quality Assessment,* DAMA UK, October 2013, www.damauk.org/community.php?sid=7997d2df1befd7e241a39169a2c95780&communityid=1000054.

- Validity
- Accuracy
- Consistency

These aspects of data quality may be considered at a few different stages, such as in setting up a data warehouse, or when beginning the modeling effort, in order to select the variables that will be most reliable.

However, when moving from your modeling phase to the implementation and beyond, it's important to avoid assuming that your initial data assessment is sufficient to benchmark data quality for implementation. When assessing data quality at the modeling stage, your attention will be on a different set of criteria than when you consider the ongoing reliability of your data post-implementation. Therefore, if you don't make a conscious decision to check that data is suitable for implementation, you can overlook some aspects.

It is very important to know what's normal for your input variables from the start. Without sampling at the start, you won't know what you should expect to see, so you won't know when things have gone askew.

It is relatively common for machine learning texts to partially ignore exploratory data analysis as it relates to the problem of ensuring that the model trained on yesterday's data is still suitable for tomorrow's conditions. Instead, it is more common for texts to emphasize exploratory data analysis as an antecedent to data cleaning and preparation, with discussions of how to deal with missing data and possible approaches to data cleansing leading to advice on the best ways to prepare a data set for modeling.

While this is a necessary step toward building an effective model, this step should also provide a wealth of information.

Once you have implemented your model as a scoring engine fed by a stream of updates to the data set, there will be another set of issues to contend with respect to the reliability of that update process. Even the best maintained and coded systems will have some level of errors, dropouts, and discontinuities in the feed. If they are at a low enough level, they won't be alarming. However, similar to the overall problems with the data set, if you don't look, you won't find out, and what you don't know can hurt you. Therefore, you need to understand the characteristics of the data feed.

This may be considered by some as the purview of the data engineering team or of the database maintainers. However, while I don't believe in the unicorn data scientist, and neither should you, and you should take help in maintaining your databases as much as your organization allows it, I believe that data governance is too important for the data science team to let themselves be out of the loop.

Therefore, it is important to have enough of an understanding of data governance best practice to ensure that you contribute to the process.

Similarly, you should expect to lead algorithm governance, a discipline in its own right.

An additional factor often associated with model and data governance is modeling bias, especially when the model bias manifests as bias against particular groups of people—ethnic subgroups or socioeconomic subgroups being some of the more distinctive ones. If a model is designed to make decisions that impact on people, these effects are an ever-present risk and one that is difficult to identify without the involvement of humans who can adjudicate on the effect of the model.

The general activity of ensuring your model is still operating within its intended parameters intuitively requires a strong understanding of what those parameters were. This understanding encompasses both the data itself in terms of statistical measures of central tendency, spread, skewness, etc. and to the data collection process. That is, changes to the rate of data capture could be an indicator that there has been a change to the underlying process.

For example, if the feed for a particular input variable sees a different pattern of updates, or the volume of data suddenly increases or decreases, these could well be a sign that some aspect of the data collection has changed, potentially meaning that the data's underlying quality or meaning may have changed, even if it is not immediately apparent based on the data's direct sampling statistics.

Another consideration is what happens if there is a temporary interruption to the data feed or a partial interruption where some variables are interrupted but other variables continue to be updated. Does this mean your model's scoring output is incorrect? At what point can you identify that there has been an interruption?

The approach to understanding the situation before you began modeling can also be used to search for bias in models. For example, it has been observed a number of times that models of crime patterns reflect the same racial or other biases of arresting officers, so that a model trained on data where blacks are over-represented in the arrest rate is likely to lead to blacks continuing to be over-represented at the same rate.

Knowing what the data looks like, compared to how it is expected to look or how it ought to look are crucial for deciding whether the results it produces are to be relied upon.

Auditing Models

One of the attributes of a reliable model I mentioned earlier in this chapter was auditability. Auditing is simply a formal process of checking that your models perform as expected, have been implemented as expected, and were originally developed as expected.

Therefore, auditing the machine learning system you implement is, therefore, the most systematic way you can ensure that you consciously look for problems, find them, and document potential solutions. This realization is becoming more widespread, and as new pressures are applied to machine learning systems to be more transparent and reliable, the idea of formally auditing those systems is becoming more widespread.[8]

Although the word is often associated with accounting or tax, in fact, audits can occur in many contexts, and an audit of a machine learning system will intuitively have more in common with audits which occur in IT or similar contexts.

A big part of the need for a formal process is to ensure that nothing is left out of the review, as the system that needs auditing consists of several parts that create small gaps that can result in communication being lost. At the same time, although an audit is a defined process with steps that are to some extent predetermined, when an audit is performed by a human there is an opportunity to drill deeper into any areas that appear suspect.

The combination of predetermined steps that ensure that the right areas are covered along with the ability to use intuition to guide the search toward difficult areas is a strong way to deal with the difficulty of finding times when your model is biased.

Another key advantage of performing an audit, whether it uses an internal or external auditor is that you can use it to enhance the reputation of your model and its implementation. This an important concern when your goal is partly to improve users' trust in your model.

To complete an audit, you need a framework to audit against, usually, one that is either formally or informally recognized as a standard. Data Science does not have a great deal of choice in this area, but at least one author has suggested auditing Data Science projects against CRISP-DM.[9]

The advantage of the CRISP-DM process in this context is that you can use the subheadings with the framework to generate questions and discussion on how the data science implementation you are reviewing behaves against each of those points. We have seen CRISP-DM in Chapter 2, but to emphasize those subheadings:

- Business Understanding
- Data Understanding

---

[8]James Guszcza, Iyad Rahwan, Will Bible, Manuel Cebrian, and Vic Katyal, "Why We Need to Audit Algorithms," *Harvard Business Review*, November 28, 2018, https://hbr.org/2018/11/why-we-need-to-audit-algorithms.
[9]Andrew Clark, "The Machine Learning Audit—CRISP-DM Framework," *ISACA Journal*, Vol.1,2018,www.isaca.org/Journal/archives/2018/Volume-1/Pages/the-machine-learning-audit-crisp-dm-framework.aspx.

- Data Preparation
- Modeling
- Evaluation
- Deployment[10]

Each of these subheadings is a natural topic to review and validate a data science implementation against. In particular, although it can sometimes be tempting to decide that if a model achieves well against its accuracy metric, the inclusion of "Business Understanding" and "Deployment" as topic to audit against should mean that, on the one hand, the model really does provide data that answers the client's business problem and, on the other hand, the version that is implemented really does provide the client with the results they require.

Of course, there is no existing mandatory standard or even universally accepted standard means that you are not tied to CRISP-DM if it doesn't meet your needs. You can vary how deep you want to go according to your needs, and you can also add to the list if you think something is missing.

Certainly, you will find that if you approach a large firm to validate a model, they will often have developed their own framework that usually covers a similar but not identical list of concerns to CRISP-DM.

Model Risk Assessment

Another idea coming from quality assurance that you can adapt to the problem of ensuring your model achieves the hoped-for results is performing a risk assessment before implementation. This concept is particularly prevalent in the automotive manufacturing sector, but at least partly thanks to Six Sigma that has spread to other contexts.

The formal tool for doing this is called Failure Mode and Effects Analysis, usually shortened to FMEA. Ultimately, this is a process of guided brainstorming for things that can go wrong and their consequences, and you can adopt the level of formality that suits your situation.

---

[10]Pete Chapman, Julian Clinton, Randy Kerber, Thomas Khabaza, Thomas Reinartz, Colin Shearer, and Rüdiger Wirth, *CRISP-DM 1.0*, www.the-modeling-agency.com/crisp-dm.pdf.

It's possible to conduct an FMEA at variable levels of thoroughness, with different groups of possible stakeholders, and a variety of different prompts. However, a few core activities are common, no matter what the scenario:[11]

- Identify possible failure modes—ways that the process could fail
- Identify the possible outcomes of those failures
- Quantify the severity of those outcomes
- Develop a response plan for the most important failures based on the severity of their outcomes
- Document the outcomes of the process

In an environment which values formal processes and documentation, quantifying the severity is often done by calculating a risk priority number. The risk priority number is calculated by multiplying figures given for the severity of the problem, how likely it is to occur, and the probability of detection.

However, it is possible to adapt the philosophy of this approach to your situation in a less labor-intensive way by using the group to brainstorm for an overall priority score, for example, out of ten. By taking that approach, you can get the assurance that an FMEA can provide in detecting problems before they occur without taking the whole bureaucracy of the process. That way you won't let the perfect be the enemy of the good.

The usual outcome for this process is to develop a control plan, which matches particular hazards to a preventative measure. In a manufacturing context that might mean checking machine settings if the outputs change but prior to being forced to reject the product or analyzing the raw materials. In the case of a machine learning model, it is more likely to mean a sufficient change to an important input variable's data feed triggering an investigation of the data source, which could be a vendor, a collection point, or a sensor.

The use of FMEA or another tool with a similar philosophy, to look for weak points in a process or implementation, has enjoyed considerable success in its original context in the manufacturing industry. Without a process to systematically look for these weaknesses, product launches in manufacturing would very frequently lead to defective products being shipped to the customer, and manufacturers being constantly surprised by customer reports of defects.

---

[11]Roger W. Berger, Donald W. Benbow, Ahmad K. Elshennay, and Walker HF, *The Certified Quality Engineer Handbook* (Milwaukee, WI: ASQ Quality Press, 2002).

In the current environment, where users may be skeptical about data science models due to reports of poor performance in the media, or prior experiences of their own, adopting those same tools can help to ensure that your final implementation provides a positive experience that causes users to trust data science again.

Model Maintenance

Any statistical or machine learning model will experience a loss of performance over time as relationships alter. Sometimes this happens very suddenly, such as happened to many credit default models during the GFC. Other times the degradation takes place over a longer period, and can almost be predicted by someone watching the trend.

What drives the degradation? First of all, no matter how careful you were, to some degree your model fit on noise or on latent factors, which is to say it was wrong, to begin with, and some of your accuracy was due to random chance.

Intuitively, a point that emerges from those two examples is that models which are dependent on human behavior may be especially susceptible to degradation, whereas models that relate more closely to physical processes in some sense may have some additional stability. It follows in turn that a key ally in understanding how much of a risk this is for your model and over what timeframe will be your subject matter expert, and in most cases a regular schedule of model review and retrain will be developed.

At the same time, you will likely want to use what your data is telling you, so you'll need methods to determine whether the newly arrived input data has changed. This is particularly the case for circumstances which change quickly.

In the case of input variables where the data points have a high degree of independence, control charts, as used in statistical process control (SPC), could be used to detect changes to the process.

There are many guides to the use of these charts, both in print and online, and they have been successfully used for many years. Their common element is that measurements from a process are plotted sequentially on a chart with a centerline at the mean (or other appropriate process average) and upper and lower lines to represent the usual process range. Accordingly, it is easy to establish when the process has changed either its range or its average result.

However, especially for attribute or categorical data, methods developed for use on relatively small data can give problematic results when used on much larger quantities of data.

Some care is still required in setting up the sampling regimen for continuous data—and note that it is not necessary to use the complete data collected every day to check whether the input variable's process maintains the characteristics it had when the model was implemented, just that it is large enough to be representative.

Systematic Data Monitoring

As long as you have a clear understanding of normal and abnormal data in your context, data monitoring is an area that offers a lot of opportunities for automation, at least as far as there are clear ways to be systematic.

An intuitive approach is to adapt the principles of quality control monitoring where statistical charts are used to detect when there has been a change to a series of data being collected as a way of alerting a responsible person to check conditions and possibly act.

Control charts originated in the context of manufacturing, where an operator measuring an aspect of the product being made can quickly assess whether the underlying process has changed. Control charts come in a variety of formats to suit the kind of data they are being applied to, but two of the most common forms are the x-bar and R (x-bar being the mean and R being the range) and the c-chart (for countable events).

SPC control charts are interpreted in the context of a predetermined set of rules that identify when the process has changed—these include things like seeing seven or more consecutive points above the mean or seven or more consecutive points rising or decreasing, although there can be a small amount of variation on which rules to use.[12]

The principles of creating and interpreting quality control charts generalize well to a variety of contexts, and workers in data quality have increasingly recognized how they can be used to detect occasions when the quality of data might have unexpectedly been compromised.[13] They also lend themselves relatively well to automation, although a human reading the chart after an automated rule has alerted them to an issue is in the best position to decide what has actually occurred.

Specific charts have now been proposed or adapted for use with data streams.[14] Researchers have noted that the statistical properties of data streams don't always follow a normal distribution as assumed for some basic types of

[12]Douglas C. Montgomery, *Introduction to Statistical Quality Control* (New York: Wiley, 1997).
[13]Rajesh Jugulum. *Competing With High Quality Data* (New York: Wiley, 2016).
[14]Miaomiao Yu, Chungjie Wu, and Fujee Tsung, "Monitoring the Data Quality of Data Streams Using a Two Step Control Scheme," *IISE Transactions*, October 2018, DOI: 10.1080/24725854.2018.1530487.

statistical control chart, especially when attempting to measure the six dimensions of data quality we first listed earlier.

Remember that the more common way to use a control chart in most contexts is to trigger an investigation, even if it's very brief, to decide the best course of action, rather than to impose an automatic adjustment. This makes sense as these are tools to identify that something unusual has occurred, but have very limited or even zero ability to identify what the unusual thing is.

Specific control charts have been developed to identify when processes stopped operating as expected—the cumulative sum (usually abbreviated to "cusum") chart is a prominent example.

Knowing when a process stopped behaving correctly is an important step toward discovering what happened—if you have an external data provider, providing a date when something changed means they are far more likely to be able to suggest a possible change. Likewise, when you capture your own data, being able to go back to a specific date will be very helpful when you need to identify the cause of a change.

At the same time, in an ideal world, you would have identified at least some of the likely causes of disturbances during the planning stage, such as through the FMEA or control plan activities. If you have, you will have a head's start on where to look for the cause, and possibly a default corrective action to take (although you shouldn't expect at the earlier stage to have anticipated everything that can happen).

The processes of auditing your machine learning system, assessing the risk of poor outcomes, and monitoring its inputs and outputs are the best measures you can take to ensure that your project delivers its expected results. The risk of not performing these actions is similar to the risk of building a car that doesn't turn on when activating the ignition, or, possibly worse, does turn on but fails to steer or brake correctly.

Summary

For users to trust the models you create, they need to be both credible and reliable. This chapter focused on keeping models reliable. Although there are a number of aspects of reliability, there is a common thread among them that if you don't look for problems, you run the risk of not discovering them until your users do, and therefore after you have lost your users' trust.

At the moment, there is extra attention on some of the ways that models have problems because of increased awareness that models can be biased in the plain language meaning—they can discriminate against minorities or otherwise work to unfairly disadvantage subgroups within the population being modeled.

An intuitive way to detect problems in the wide sense of the word is to audit the model at key milestones in its life cycle. Intuitive examples include immediately prior to implementation and on its first anniversary.

By conducting an audit, you are able to look beyond some of the purely technical attributes (although they will usually be included) to questions of whether the model achieves its business objectives, whether the infrastructure into which it is implemented ensures it maintains the expected performance, and whether there are unintended consequences.

You can also guard against potential side effects and unexpected result by performing a preimplementation risk analysis, for example via FMEA. This is a common tool used for quality assurance that attempts to anticipate problems before they occur, and put in place either measures to prevent the problem or mitigate the effects.

Complementary to both of the ideas mentioned earlier is monitoring both the incoming data feeds and your model's results. To ensure that your model achieves the accuracy you expect, the incoming data and the results should have a distribution close to their distributions when you developed and validated the model. Moreover, spikes or dips in the rate of incoming data could indicate a change to the underlying data itself, which could compromise the model's results.

Statistical process control charts have been specifically designed to monitor processes in order to detect when they stop behaving "as usual" or "as expected." By establishing the usual statistical range of the process, they mean a quick visual check of your process can establish whether or not your process is operating as usual, and assist in establishing when it stopped operating as usual.

These actions which apply quality assurance thinking to data science models can be expected to increase the extent to which your system behaves as expected and intended. They will allow you and your users to be assured that the claims you make will be fulfilled by the model.

In Chapter 6, we will examine how to communicate those claims to the best advantage.

Reliable Models Checklist

- Have you conducted a risk assessment prior to implementation?
- Have you set up monitoring of the incoming data quality and its distribution, and likewise for the outputs?

- Have you developed a plan for the actions or investigations that will take place if the data turns out to be outside your limits?

- Do you know which aspects of data quality are relevant to your implementation? How can you detect if their behavior has changed?

- Have you taken stock of the outcomes of your model, and attempted to discover unexpected and unintended consequences of the way that you have organized your model?

- Do you have an agreement with the model users on how often the model will be checked and retrained? How is that agreement documented?

Promoting Your Data Science Work

The previous chapters have seen us land the opportunity to do a useful data science project, confirm the customer's willingness to implement the results, and ensure that the desired results will be achieved. In this chapter, we will look at how to ensure that your efforts up to this point are recognized in your organization and beyond. That recognition can be an important capital for landing and taking on more exciting work.

We have seen the importance of getting people's attention to be allowed to begin and implement projects. You might be forgiven for thinking that that's enough—you convinced someone to let you build something for them and delivered the goods.

Unfortunately, it's only part of the story. You can't rely on people to recognize the benefits they've been given, and you can't rely on word of mouth to let them know. You have to take responsibility for people understanding the value of what you've done.

© Robert de Graaf 2019
R. de Graaf, *Managing Your Data Science Projects*,
https://doi.org/10.1007/978-1-4842-4907-9_6

There are a few different modes to this. One mode is recording what happened inside your own organization and letting people know it happened. Another is communicating to people outside your organization or people who are otherwise new to you.

People often think of communication with the world outside their organization first, as that's what promotes sales. That shouldn't mean that if you are primarily in the position of trying to improve an organization from the inside, you shouldn't be communicating in a very similar way. In fact, we will see that the structure and content transfer from one situation to the other quite well.

This may be more difficult than contacting people in your own organization. External people are less likely to realize you exist, to begin with, and have less inclination to assume that you are doing something that is of value to them. The idea behind the next tool we will discuss is to grab their attention by offering some information for free. The tool that has been developed to promote experts in this way is the whitepaper. It ought to be easier to get the word out to people in your own organization. They work within the same four walls that you do, so you have some ability to access them face to face. For this situation, the best thing to happen is that you can give a talk on your work directly, and we'll discuss some ways to get the most benefit from that later in this chapter.

Data Science Whitepapers

A key way to let people know what you've done is to write a whitepaper. A whitepaper is a marketing document, which aims to showcase the author's expertise in a particular area.

When writing a whitepaper, typically the author will try to either explain how they solved a problem with their expertise or teach some basic aspects in their field, with the aim of helping the reader understand when it's time to call the experts. Hence, a tradesperson might share some tips around some very small jobs, leading up to the point that the reader should call in the professionals.

There are a great many guides to writing whitepapers throughout the Internet, often including a guide to structure. They come in enough variety that you can choose the one that fits your needs the best, so have a look at a few and choose one that makes sense in your own head. Two, to start you off are the comprehensive guide by the Content Factor[1] (a whitepaper itself, hosted at a web site with other good example whitepapers) and the guide from Foleon. com,[2] which has some pointers on how to distribute your whitepaper that aren't found in every guide.

---

[1]"Eight Rules for Creating Great White Papers," The Content Factor, accessed April 6, 2019, from www.contentfactor.com/free-whitepapers/.
[2]"2019 Ultimate Guide: Writing a White Paper," Foleon, accessed April 6, 2019, from www.foleon.com/topics/how-to-write-and-format-a-white-paper.

In the case of data science, though, there is a twist, which is that usually the author is using their data science expertise to solve a problem in an area where the reader is an expert, whereas normally the writer is an expert in the domain and the reader is not. This has a small but noticeable effect on the way that the document needs to be structured and how to approach the audience.

As a result, your first task will be to establish credentials in the area of reader's problem domain, and as you are unlikely to have higher qualifications or experience in that area than the reader, straightforwardly offering up your own credentials is unlikely to succeed. Instead, the best path forward is likely the "show, don't tell" approach, often seen in creative writing classes.

In this context, it refers to allowing the reader to see your characters in action and their story unfold, rather than writing out their traits or outlining the plot. In this context it means explaining the domain problem you worked on in a way that leaves no doubt of its importance to the field.

You wouldn't be working on it if a solution wasn't valuable, so explain where the value lies—many times it will take the form of this problem being a roadblock for a bigger target. Overall, demonstrating you understand how the problem affects their business allows you to win the audience over.

In many ways, this process within the paper is simply a recapitulation of the journey towards establishing trust I've presented in earlier chapters. The difference is that this time you don't have the benefit of being in the room with your audience to begin a two-way conversation—you have to anticipate a little of the audience's possible reaction to ensure you get there.

Once you've established the problem, the next step in the story will be how you solved it. In the context of data science, two tools will commonly be needed to obtain a solution—an adequate data set ("adequate" because most data sets fall far short of our "ideal data set") and suitable analysis tools.

Given that so many data science tools are open source, there is a reasonable chance that the data set—if not in its original state, often after the cleaning and preprocessing you've performed on it—represents an advantage over competitors.

Hence, mentioning either the way the data was obtained or cleaned may be useful to further establish credibility. This is especially the case if you used advice from subject matter experts to improve the preprocessing process, for example, if there was a reason for missing data relating to the collection process that determined how those missing data were treated.

When discussing the algorithm used, it's not just a question of correctly tailoring the discussion to a nontechnical audience but also a question of pacing. To maintain your readers' attention, the whitepaper needs to have the feel and pacing of an unfolding story; too much detail on how the algorithm works and how you did it will slow the pacing and put off the reader.

Crucially, it is not necessary for the reader to come away with a complete understanding of the algorithm used to get your message across. It is almost more true to say that any description of the algorithm provides more color and interest than it provides a true explanation of how the algorithm works.

Applying your algorithm to data represents the second act in your three-act story, the first act being understanding the problem and the data. Here, the solution itself may not be the selling point, as important as it is. When you're implementing something similar to a predictive model, the selling point will often be what you observed about your data along the way—an extra lesson about the way the variables interact with each other or a surprise about which variables are the most influential or the shape of the relationship.

Although the focus with whitepapers is almost always on external readers, there is room for whitepapers that are aimed at internal users or that don't make a distinction. As an engineer, I worked for a manager who had the team maintain a library of whitepapers by another name on a variety of issues. They were good both for distributing directly to the customer and for keeping a variety of customer-facing staff informed.

Your whitepaper will make people remember you and think of you as someone who can be useful in their field if you frame it correctly. One of the biggest barriers to acceptance of data science solutions is going to be a feeling that data science is usurping expert knowledge—the whitepaper represents a golden opportunity to show that data science is not a usurper, but is complementary to expert knowledge.

Talking About Your Work

For an internal audience, you probably have more access to your audience, so you aren't restricted to using a whitepaper to promote yourself to people within your organization (although we will see later that there are still times when that might be useful).

The best way to spread the word about what you have achieved is to do it face to face with a talk or a presentation. Back in Chapter 3, we looked at some of the things that make a successful data science talk from the point of view of persuading an audience.

It was clear in that context that persuasion was the main goal. You were trying to win over an audience who was going to decide whether or not to proceed with your project. You may believe that you've passed that point and can go easy on the sales pitch and proceed directly to dispensing information.

The danger for a lot of data scientists is to assume that the objective in a data science presentation is solely or largely to deliver information. This can be true if you are delivering a presentation to an audience of other data scientists to explain a technical point. However, this is not likely to be the most frequent or the most important scenario in which you find yourself giving a presentation.

The more frequent scenario is that you need to persuade people that what you are doing is a good idea, or convince them that your work has had a positive effect on the organization.

The reasoning comes back to a point that is frequently made in guides to preparing presentations, whether they come in five, six, or eight steps (a matter of personal taste—just choose the one that makes the most sense to you, similar to guides to whitepapers)—one of the first steps when preparing a presentation is to consider your audience.

As a data scientist speaking to an audience that contains non-data scientists, you can't assume that your audience accepts that your work has value and you can't assume that your audience has the same concept of what the value of your work is that you want them to have.

Even after the work is complete, you need to continue to sell the benefits. You also need to continue to avoid adding in technical details your audience will find extraneous.

Your tale of how difficult it was to speed up your algorithm considering the ancient machine and unsuitable operation system it was running won't excite a non-data scientist audience. Your non-data scientist audience won't care how impressed other data scientists were with the technical brilliance you displayed implementing a new kind of algorithm in a difficult to use coding platform—this audience won't understand why it's impressive and you will lose their attention in those sections.

Instead, they will care about how your innovation will reduce the time they spend doing things or how it helps ensure that what they spend their time on pays off. By sticking close to these attributes of your project, you will ensure that your work is remembered throughout the organization. Otherwise, you may find yourself relegated to explaining to external consultants what the different tables in your company's data warehouse contain, so they can earn a huge multiple of your salary to do something you could do in your sleep.

This time, however, the benefits have either been realized or are about to be realized. Hence, for this part of the process it is important to have reviewed whether or not those benefits have been realized or are likely to be.

As much as possible stick to undisputed gains, otherwise you will run the risk of being challenged in your own meeting. If you are challenged successfully, you run the risk of losing some of your license for future efforts, which obviously defeats the purpose of having this sort of meeting.

In most cases, the gain will be big enough to not require embellishment, so you avoid making claims that can't be substantiated or will antagonize your audience. Also avoid the temptation to oversell by referring to gains that haven't occurred yet, especially gains that require additional rounds of work. Just stick with what's happened so far.

There is a sweet spot on how often you do this. People are happy to take a few minutes out of their day to learn about the rest of their organization and about initiatives that make their jobs easier, but there is a frequency at which the exercise becomes a little too routine. Four or five times a year is probably the upper limit.

However, less than about two times and you'll end up being forgotten, so make a commitment to yourself to get in front of those internal audiences by diarizing times to look for work within your department that is suitable for sharing with the rest of your organization.

If your organization is large, though, this could actually translate to more than two or three talks through the year as you will likely present to different groups at different times.

Presenting to the Outside World

A lot of data scientists do present their work to groups, to people outside their own organizations, for example to data science meetup groups, or to similar groups in their area. The reason is often to promote the data science team of your company as a great place to work that is doing interesting work.

More than that, by creating a presentation based on your work, thinking through how to engage an audience with that material, and therefore working out what about it would make an audience engaged, you will work out what about your work was important.

You can enhance that last benefit further by breaking free of the tendency to think about work in terms of specific projects and thinking about common lessons that apply across multiple projects. These could be technical lessons, for example, best practices that apply to particular tools, or more human-centered, such as the best way to talk to customers from a particular background or in a particular situation.

It also gives you a great chance to get a free validation of your model via questions from the audience or people who ask you things afterward. Although people are likely to be polite and encouraging, when they have questions you can't easily answer, you will know they have found a hole.

The other side benefit is that as you will want to rehearse your talk with a friendly audience, most likely made up of people from inside your team, it's an additional opportunity to talk to those people about what you've been working on or what the team has been working on, and an especially good opportunity to discuss the team's work outside of the narrow goals of individual projects.

At the same time, there will only be a limited number of meetups, so the opportunities to speak will also be finite. Fortunately, you can achieve many of the same benefits by blogging, especially if you publish with a site like Medium.com, where there's a decent-sized audience.

Even if you don't use a site with a large audience that is generous with comments, the process of deciding the best projects or lessons to choose, and then explaining them from scratch to audiences who should be assumed to know nothing about your organization will help you reconsider what you are doing and find new ways in which your work is exciting.

Lastly, in both of these situations, audiences of other data scientists are the people who will give kudos for technical achievements you can't get from a lay audience, as mentioned in the previous section. If you want feedback on your novel technical solution, these are the key avenues to find it.

Making History

One of the great things about being a data scientist is getting to try a lot of different ways to solve problems. Naturally, many of these attempts will be glorious failures, where the intended problem is not solved, but something is learned that can be used elsewhere.

Many will also be straightforward failures, where all you learned is that the proposed technique is not the right approach for that problem, or, at least, that the proposed solution requires too much effort to justify the payoff. These are important lessons to learn if your organization doesn't want to try the same unsuitable approach on each problem every other year.

Therefore, you ought to be as proud of your failures as you are of your successes if you are to ensure that people don't attempt your failed pathways again and again. Strange as it may sound if you are in the middle of working on something that looks like it won't turn out right, the thought of the organization repeating your mistakes is more embarrassing than letting your co-workers know about them as they occurred.

To ensure that others don't follow in your footsteps when you'd rather they didn't requires you to be forthright about what has worked and what hasn't. At the same time, as most of the time you won't be able to predict precisely who it is that's going to repeat your mistakes, you need to keep this information in a way that future users can find it.

This is one of the key outcomes of project documentation—recording what worked and what didn't for someone in your shoes in the future, who could be you or could be someone else.

You can capture this aspect of your projects via "Lessons Learned" documentation, which should be considered a significant deliverable from any data science effort.

Effectively, these are documents where you record what was attempted, what worked, and what failed. They are different from a laboratory notebook, however, in that they are intended for a general audience, rather than just as a personal aide-memoire.

Therefore, you need to think carefully about how you will structure your account to match the intended purpose. In this context, the important thing is to cut to the chase, so that someone who wasn't around to understand the context of the project you have been working on can still easily understand the important lesson that was learned as a result. As much as possible, leave the details of the business case for doing the work out—just enough information that people understand why you looked in this area at all.

Pro Tip   Successfully creating a library of information on previous efforts that is well used can be a substantial competitive advantage. For example, the author of *The McKinsey Way*[3] says one of the great advantages of being at McKinsey is being able to access the database of work McKinsey has done on previous projects. On the other hand, in my first couple of jobs, I spent a lot of time re-establishing knowledge that had been lost, and I can attest that simply reinventing what you know has been established before doesn't make one look forward to Mondays.

The crucial part is to agree where to put the lessons learned documents, as these will be essential parts of your organization's corporate memory—as long as they can be found by anyone who needs to find them. Your organization's network may be either a blessing or a curse for doing this—just placing into the shared drive runs a strong chance of seeing them forgotten and unfindable. Using a Git repository or similar is better, and is okay for the intended audience of data scientists. The trick, though, is to avoid keeping the lessons learned documents too closely tied to the individual projects they came from.

For the non-data science component of your company, it is better to get the message out through wider channels. These could include company newsletters.

In some of these forums, you wouldn't want to refer too directly to things that haven't worked as you expected. When you want to report on a data science project that didn't go the way you expected, you need to reframe it

---

[3]Ethan M. Rasiel, *The McKinsey Way* (New York: McGraw Hill, 1999).

away from the original goal. That is, emphasize what you discovered as if it was the goal from the beginning, and let the original goal appear as a secondary goal.

Differing Audiences for Documentation

The most obvious and natural audience for lessons learned documentation is the other members of your data science team. They are obviously most able to benefit directly from the knowledge that on some particular data set a particular approach that's popular within your team doesn't work as expected, or other similar insights.

This shouldn't mean that you neglect writing documents that can be understood by a lay audience, especially for more senior management. If you fail to keep management informed on what you have learned, you run a strong risk that they will ask you to repeat work you have already done that you know won't achieve the desired result.

At the same time, senior managers and others outside the data science function are unlikely to have either the time or the inclination to wade through the details of every project to discover the most important lessons learned for them. Instead, you've got to go to them.

When you prepare the documentation for this audience, you need to ensure that it is intuitive for them and speaks directly to their need. It's fine to be local, in the sense of referring to in-house data sets, or your organization's customers or product lines with your in-house terminology, but the technical side has to resonate with their level of understanding. Don't be embarrassed to keep it very simple.

To help ensure that people who want the technical details can read them, and those who don't can avoid them, consider the structure of your document carefully. By dividing the document into sections and marking it out with clear subheadings, you can help people find the parts they want to read most easily.

Finally, keeping it as short as possible will maximize the chances that people read enough of it to read the parts you want them to read. Obviously, every extra word you add adds to the risk that your reader loses interest and stops reading.

The totality of this advice might seem very familiar. In fact, realistically, what you are doing here is really creating a whitepaper for internal circulation within your company.

The goals are actually surprisingly similar—you may not realize it initially, but half the purpose of these is to ensure that you and your data science team are thought of early as the people who can help the business for any given problem.

The crucial message here is that you can help with anything, and your answer will be useful.

The difference is going to be the length. It is extra important to keep the length of your document in check when you are writing an internal document. People are more inclined to assume there is value in an external whitepaper. This is partly because people know that the authors of external whitepapers see them as a potential source of income, and partly because there is a sense that the authors of the whitepaper are difficult to access. If you are someone they see every day or believe they could see any time they had the inclination, it will make it less likely that putting effort into reading your paper will seem worthwhile—you'd better make it a short and easy read.

The lesson is that just as you would never assume that external customers continue to see your value without you reminding them every so often, you can't assume that your internal customers automatically see how valuable you are either.

Summary

Implementation isn't the endgame. You need to ensure that others hear of your best results. You also need to ensure that you are the person who communicates what happened when projects didn't succeed so you can explain the lessons you learned.

There are both written and spoken methods to do this. To promote your work, you may wish to write a whitepaper—when written well you can attract more work very effectively this way. It is important, however, to get the balance right and certainly to ensure that you are generous toward your reader. That is, you need to give the reader useful information, and not simply promote your product.

Documenting what you achieved is also essential. It might be tempting to think that the documentation that is intended for other data scientists is the end of the story. The other data scientists on your team likely know where to look for information on previous data science projects. The other group that is important not to forget is the non-data scientists, particularly as in many cases this group includes senior people who can get you to repeat work you've already done.

Although whitepapers are seen as documents for external stakeholders, you can use simplified versions of the same structure to create internal whitepapers that do the same job within your company.

It will be more work for you, but less work for your audience if you are able to present your work to your users personally. In general, people will be pleased to hear about innovations that reduce their workload, so they are keen to come to your presentation, but make sure that the results are what you will claim they are before you set up the meeting.

Across this whole chapter, one of the key lessons was the importance of learning as much as possible from your efforts while communicating to as many people as possible what you learned. These are some of the most useful initial steps you can take to build your data science team's brand.

The next chapter builds on this idea to examine how to build behaviors that help your data science team learn more effectively and function more effectively at the same time.

Promotion Checklist

- Have you developed a whitepaper that showcases a key insight you discovered in a way that also showcases your team's capabilities?

- Did you make sure your whitepaper gives the readers information they probably don't have to build trust, and that it establishes your credibility in the relevant subject domain?

- Have you presented your work to a local meetup group, showcasing a different side of it compared to the side you showcased to your customers?

- Have you blogged about your work, presenting some of the work that you haven't been able to present at a meetup group, or presenting some of the lessons you learned only by doing a few different projects?

- Have you diarized when the next two or three times you will present the data science team's progress to others in the business will be?

Team Efficiency

Making the Best Use of Everyone You've Got

Throughout this book I've emphasized getting the best value from the work that you do, whether carefully choosing the most valuable project or by making sure that the finished version of that project is appreciated to the fullest extent by the widest possible group of users.

In Chapter 6, we talked about marketing the data science team's work to others in the organization. It could be said that we were looking at the beginnings of building a team brand.

In terms of this book, it was natural to look at the team branding first, as it flows on naturally from the projects you've already completed. But it's also reasonable to wonder what happened to making the team work well together in the first place. This chapter should address that.

Sometimes it seems that the word "team" is a punchline from a comedy about corporate life like *The Office* or *Office Space*. By contrast, rather than meaning, for example, enforced "fun" with co-workers, in this chapter I want to talk about making the best use of the people around you. This is intuitively achieved by communicating better with the people around you and finding common ways to do things.

© Robert de Graaf 2019
R. de Graaf, *Managing Your Data Science Projects*,
https://doi.org/10.1007/978-1-4842-4907-9_7

Learning from Your Work

Data scientists are often focused on technical learning, but the human element should not be neglected. Throughout this book we have discussed different methods for ensuring that the human element is not neglected in terms of communicating with the people around a data science team more effectively. However, we haven't talked about how to work together as a team for the best results explicitly.

At the same time, we should take note of some of the indirect ways of improving team efficiency that we have touched upon. Some of the most important of these were in the last chapter, where we discussed ways of promoting the lessons learned from your data science process. It is certainly the case that ensuring everyone around you learns from your work as best as possible is one of the most important aspects of working effectively in a team.

However, while it wasn't said outright, the kind of learning that was going on as a result of these projects was usually technical learning. Implicit in our discussion was that most of what we were promoting was the direct result of your research or analysis.

However, it can be useful to make a point of noticing the other aspects of what you are trying to achieve—that is the human elements of what you have been doing. If you don't set out to notice these kinds of lessons at the beginning of your process, you are likely to fail to take time to recognize when you have learned from a particular project.

To illustrate what I mean by this, consider the life cycle of projects as discussed throughout this book. Throughout this book I've referred to the way projects begin with a customer or client who has a problem through a process of understanding the problem correctly, proposing and implementing a solution, and then documenting and promoting the solution that has been implemented.

In Chapter 6, when we talked about promoting the data science team by sharing what you learned along the way, you could have reasonably inferred that what I was talking about was the technical findings that come directly from the data analysis and modeling process or otherwise from the process of attempting to implement the solution. Indeed those are the right things to share with the rest of the organization when you are trying to build a brand for the data science team.

However, at least within the data science team itself, the lessons you learn about how to talk to certain people or groups of people, or a new way to write a great whitepaper, are just as significant and useful as either the direct results of analysis or a technical lesson, such as a new way to prepare some type of variable.

Unfortunately, these lessons are less often captured in formal documentation or presented back to the organization via training sessions compared to technical lessons. There could be a few different reasons for this situation, but it's likely that the perception that creating documentation and presentations on human-centered issues is seen as more difficult, especially by technically oriented people.

If you are an Agile team that does regular retrospectives, you've already got a regular process that exists in part to make sure certain kinds of undocumented communication occur. The danger can be that sometimes the focus on projects means that some of the biggest lessons can be missed.

The set up for a retrospective meeting is straightforward. You look back at recent activity and list what went well and what went wrong. In an Agile context, the recent activity often means during the last sprint, you don't have to follow an Agile workflow in order to do retrospectives. Though Agile was the first to formalize it with a name and the idea has become significantly more popular as a result of Agile, it's a universally good idea for everyone.

A critical reason that retrospectives sometimes don't deliver is that data scientists (or software developers) aren't naturally comfortable talking about the human side of delivering projects and often find clever ways to turn discussions meant to be about the human side into technical discussions.

A common way that things can go awry in this way is that people overuse the project or Agile terminology so that a discussion that ought to be about a human issue such as poor communication instead remains fixed, or drifts into a discussion about the technical outcome. For example, where a communication problem has resulted in someone receiving the wrong information, leading to a technical problem, the communication problem is the root cause of the problem but can be overlooked, and a thorough discussion of the technical consequences can be substituted for a more productive discussion of how the communication fell apart.

If you aren't in an Agile environment and therefore don't have retrospectives, or are in an Agile environment and haven't adopted them yet, it doesn't mean that you've missed out on having a retrospective. A retrospective doesn't depend on having an Agile environment. You might even be able to have a more effective retrospective if you aren't tethered to the Agile terminology for the reasons mentioned earlier.

The ways of holding a retrospective that you can find in Internet guides are effectively just the base you can use to make a retrospective that works for you. In each case, you are essentially presented with different ways to facilitate a somewhat guided brainstorming session. Within that context, you have a lot of room to guide the brainstorming where you think it needs to go the most.

Don't be satisfied with retrospective by numbers. That is, if you ask the team to come up with the things that worked and the things that didn't, and after a little bit between the plus column and the minus column, all you've got is the same old stuff you had last time couched in safe Agile jargon or the safe jargon of the last training session your company paid for—reject it and ask for more.

If it's not quite as bad as that, but the only problems are purely technical, give them a few prompts for the human side of things. It could work the other way too—if the problems are too much on the human side (which could well mean that the retrospective has descended into a simple blame game).

The retrospective, in fact, is an expandable format, and similar to a master stock or 12 bar blues, it can be done in different ways to suit what's needed by the people using it.

More than anything else, retrospective is a platform for the most important role a manager can play—the manager as a coach, where the word coach itself is really another word for teacher.

In fact, although the idea of coaching can sometimes conjure up "official" or company-mandated one-on-one coaching sessions, the team coaching session can sometimes be far more effective. Consider sports teams (as much as the comparison to business teams is over-used). A huge amount of the interaction and work done by a coach is done with the team as a group rather than one on one. There is a huge opportunity to improve the team by using the sessions as opportunities for group coaching, most obviously by identifying behavior you want the whole team to adopt.

A secondary benefit to taking a conscious decision to lead the discussion in a retrospective is that you can call out for compliments examples of people improving the team by doing "glue work." "Glue work" has been loosely defined as work that is essential for team success, but not measured by the organization's standard metrics.[1] This sort of work can easily go unnoticed, and productive team members can go without receiving the due credit for the effort they apply to increasing the whole team's productivity.

This may be a different way of doing things for some whose instinct is to hang back and let the group's thoughts flow. There is a time for this, but there is also a time for ensuring that the right issues are not just discussed but that the discussions result in practical suggestions.

Therefore, there is room for someone facilitating a retrospective discussion to join in and guide the discussion toward the most important and relevant issues. Not only that, but room to challenge what comes out of the discussion in order to ensure that what's being decided is practical.

---

[1] Tanya Reilly, "Being Glue," accessed on April 8, 2019, from www.slideshare.net/TanyaReilly/being-glue.

The end goal, of course, is that what you discover through the retrospective can be applied to what you do to change your practices to get better results. When you do that, you will want to ensure that the new practices are used as often by your team as practical. That means that you need to find ways to standardize what you do.

A Shared Way of Doing Things

One of the most common lessons for improving a team's effectiveness is to have a common purpose that is understood, as far as possible, the same way by all the members of the group. This can be challenging for data scientists given the lack of an agreed definition of what a data scientist is to begin with. However, within a specific organization, you have at least some chance of being able to establish what a data scientist is in your immediate context.

Even at the practical level of a shared understanding of data science practices, the diverse range of backgrounds that data scientists may have creates a heightened need to ensure everyone in the team shares the same understanding of frequently used terminology, and the same overall approach.

Intuitively, the best way to share a vision is to create it together. Many guides to team cohesion suggest brainstorming the team vision together. We covered some of this in Chapter 1 when we discussed creating a team mission.

However, the practical side still needs attention. In many industries, for example commonly across the manufacturing industry, highly standardized processes are created and enforced from the top of organizations down. It often happens that they are then resented by operators who are expected to use them.

The situation is different in relation to a data science team. These standard processes are only meant to be used by the relatively small number of people in a data science team. The relatively small number of people also means that, unlike the case that often applies to a large manufacturer, it is very practical to choose the standard practices as a team.

The key advantage of being able to standardize a process is that it reduces variability. In a manufacturing process, several other advantages flow on from that, but in our data science context one that is useful is predictability. By having a predictable process you know what you are going to get and how long it takes to get it. These advantages are highly important again for the process of winning trust—the ability to be predictable means that you can make promises knowing you can keep them.

Note that you don't need to be limited by the meaning of the word standardization. That is, it can be wrongly assumed that standardization simply means creating a "black letter" process that everyone follows the same way. However, there are ways to standardize that don't take that approach.

Consider, for example, the Agile Manifesto,[2] which is expressed as a series of preferences, rather than as predetermined choices. This idea can be extended to other areas to mean "try this first"—for example, you could develop a guideline in modeling that you always try logistic regression first, and then move toward more complex and less transparent models.

Another way to have "soft" standardization is to create boundaries. To use a similar example to the previous one, you could have a rule that for a certain class of problems you will never use k-nearest neighbors (or some other algorithm that you have that doesn't produce good results for your typical kinds of data).

Other ways that you can standardize effectively within a data science environment might include the following:

- **Standard definitions of target variables:** For example, do you have a standard place to start when considering targets based on time windows that makes sense for your organization?

- **Standard terminology:** Does Jill say independent variable and Joe say input?

- **Standard Tools:** You've probably decided on a standard platform/language, for example, R or Python or a commercial package—but if you've gone for R or Python have you standardized on preferred libraries for particular common tasks?

All of these things will stick more easily if you decide them as a group. It's also a great standard response to things that come up regularly at retrospectives to put the question of "Do we need to standardize on that issue?" That way you've got a live example in front of you.

Standardization is often more honored in the breach than the observance—people agree that it's generally a good thing but don't do it because they have an overly stereotypical idea of what it is or how to do it.

If you break away from that stereotype, you can open the door to being able to standardize practices within your team in a way that you can control and that works for you and your team.

---

[2]"Manifesto for Agile Software Development," April 4, 2019, from https:agilemanifesto.org.

The Skills Your Team Needs

Data scientists are almost obsessed with the skills they need. This is probably because of the continued ambiguity on what a data scientist actually is. If a data scientist is, as they say, "someone who can code better than a statistician while knowing more about statistics than a coder," where does the need to learn on those two fronts end?

This notion—sometimes referred to as the unicorn data scientist—often appears to be built on the assumption that a data science team works independently of other teams. Therefore, if they need to provision a database, they will end up doing it themselves. If they need to build a UI, they will end up doing it themselves.

This might be true sometimes for prototyping or developing a proof of concept, but it's more likely in companies that are larger than the data science team itself that there will be people whose specific task it is to do these things. For example, they might exist under a variety of names, but there is likely to be someone whose job is a descendant of database administrator, and they'll often have existed in your organization for a long time before the data science function.

Where they do exist, you have an opportunity to move that work outside of the data science team, which simplifies the skills you need to maintain. Don't worry, there will still be a long list of things that can only be done within the data science team. In fact, it's exactly because that list is lengthy that you need to be careful to avoid doing things that you don't have to do.

The knitting you need to focus on is the part of the job that can't easily be done by other people, or at least needs to be understood within the data science team to ensure the best results.

An example of the first of these is model evaluation—those skills just won't be found anywhere else apart from in the data science team, so they had better exist there and be performed well.

On the other hand, although an understanding of the business will obviously be found elsewhere, and frequently be better developed than what's available in the data science, it can't be outsourced the same way that building an ETL can be outsourced—a sufficient level of business understanding is essential within the data science team.

Therefore, when developing a skills inventory, you need to develop it at two levels. One level is internally for your team and the other is your team in comparison to the rest of the organization.

Also consider that data science job ads are often framed in terms of a list of tools mastered or in terms of particular skills areas. Technical skills are often mastered relatively easily by people with the right way of thinking. Harder to pick up easily are those mindsets themselves.

As an example, you could divide people into "builders" and "analysts"— people who want to work on building data products vs. people who want to analyze data to understand how it applies to a problem. These are very different mindsets. Another different kind of role is a "spanner."[3] Spanners, in some ways, are what many people think of data scientists—they are people who span the gaps between builders and analysts or between data scientists and data engineers. Again, although it is true that to succeed in this role there is a need for skills across multiple areas.

Having a strong understanding of the skills that are available throughout other areas of your organization can take some of the pressure off your data science team generally, and help you to keep your list of desirable skills manageable when you need to hire.

Summary

Data projects need data science teams to complete them, but often data scientists focus on the technical details of their projects and don't worry as much as they should about how their teams work, or even how well the data science team plays with the rest of the organization it's a part of.

There is at least some of this human side in Agile. Retrospectives are meant to capture some of the human problems, although they are sometimes considered one of the more difficult aspects of Agile to get right (although you don't need to be officially Agile to have retrospectives).

While there are many guides to performing retrospectives, a key element of ensuring you successfully discuss the most relevant human problems in your process and discover practical solutions is ensuring that the human aspects are properly discussed. When dealing with a group of people selected for technical skills, this will often require someone to lead the conversation onto the right topics.

Sharing standardized approaches and a standardized vision is also a crucial method of improving both team cohesion and team efficiency. In data science, there is arguably less standardization of training compared to a profession like medicine, so the need to take deliberate steps towards standardization is heightened.

---

[3]Wayne Eckerson, *Secrets of Analytical Leaders* (Westfield, NJ: Technics Publications, 2012).

Understanding the skills in other areas of your organizations can help you to keep the skills requirements in your own team manageable. Identify work that doesn't need to happen within your own department, so you don't need to constantly maintain those skills and you can simplify your own processes. You can also simplify the types of people that exist within your own area and therefore improve cohesion.

However, standardization doesn't have to mean the sort of strict processes that are often associated with that word. There are creative ways available to provide guidelines that don't also impose onerous restrictions on your team members' creative approaches to their work.

Team Efficiency Checklist

- Have you created a process of team retrospectives that considers the right things in terms of sharing the lessons from previous projects?

- Have you practiced ways of soliciting extra attention toward the human elements when doing retrospectives?

- Does what you decide in your retrospectives carry over to what you do in your normal day?

- Have you created standard terms for data science concepts and business concepts for use within your team, and a standardized understanding of priorities?

- Have you created a team vision with input from everyone on the team?

- Have you assessed the skills required in your data science team, considering the skills that are available elsewhere in your organization to ensure your team members develop the right skill set?

Afterword

Over the course of this book we have developed a strategy for data science projects which includes the beginning stages of uncovering what the customer really wants, through to understanding the roles that your team plays in delivering to meet that need. It is now time to take a step back and see the whole picture.

What's the difference between a data science project that succeeds and one that fails? A successful data science project begins with a clear understanding of the customer's needs and ends with results that can be understood in a usable platform, and along the way, the project leader must convince people that the project is worth doing.

Examining the process from start to finish enables you to not only understand what is needed at any particular moment in the journey, but to understand the bigger picture requires you to step back and think about how the different regions relate to each other and how to develop a common framework to apply to improving the efficacy of your data science team and your data science projects.

One word has reappeared constantly throughout this journey into better data science projects: trust. The big problem when you're trying to get someone to trust you is that it doesn't come overnight—instead, gaining someone's trust is an incremental process, similar to the stages needed to create an oil painting, from first sketch to detailed coloring.

For data scientists, trying to gain the trust of others in your organization or trying to win the trust of customers can seem very slow. Making a model is very quick. Activities like data preparation are much slower, but still not as slow as convincing someone to trust you.

© Robert de Graaf 2019
R. de Graaf, *Managing Your Data Science Projects*,
https://doi.org/10.1007/978-1-4842-4907-9_8

Fortunately, an individual project often provides many opportunities to build trust, as we have seen. There is a crucial opportunity when you first engage with someone who has a problem—you can win their trust by listening and understanding their problem, nearly to the same extent as you can by actually solving their problem.

Gaining trust means not wasting opportunities. A cookbook published around the beginning of the millennium advocated nose to tail eating.[1] The philosophy of nose to tail eating is often summed up by the author's slogan, "if you're going to kill an animal, it's only polite to eat the whole thing." In data science, if you're going to take someone's time and their data, it's only polite that you discover all the lessons that are available in doing so.

Every time someone lets you work on their problem, they are taking a risk. Repay that risk by understanding that working on someone's problem is a golden opportunity. Any project you work on gives you multiple opportunities to convince someone to trust you, even if the project doesn't make it through to final implementation.

Don't waste those opportunities, especially the opportunities that seem like you missed. After all, it's no great challenge to make the best of a project that turned out well. The trouble is that there will be projects that don't turn out well, and you still need to make the best of those.

To make the most of those missed opportunities, you need to take as wide a view as possible of what the possible lessons are. If what you learn from a data science project is how to work better with your local database administrator, there's some value there, and if the two of you now have a better working relationship, you owe it to the customer who helped you get there to find something of value in the data together.

It's only polite to make the most of what people give you.

Building trust into your models is a virtuous circle. When you work as closely as possible with your customer, you build something they want. When you build something they want, the more freely they will talk to you about their real needs, and the more likely it is that you will build something they want.

To be sure, there are exceptions, and it can happen that success against reasonable expectations is rewarded by new unreasonable expectations. Even so, it is far easier to win people's trust, and therefore ensure they give you their time by succeeding at what you do, and convincing them that you have even more to give them.

Losing trust goes the other way. If they don't trust you, the more guarded they will become, the less open they will be about what they actually want, and the more likely that what you build doesn't meet their requirements.

---

[1]Fergus Henderson, *The Whole Beast: Nose to Tail Eating* (New York: Ecco, 2004).

Another outcome is that the choice of deadlines can become less reasonable—and less flexible if changed is needed—further deepening the trust deficit. This is a vicious cycle indeed.

Far better to be in a cycle that repays you by increasing your chances of success, rather than defeats it. Moreover, if nothing else, this book should have shown that you have considerable control over the way people react to your work.

At the center of data science success is the ability of your project to solve a problem for humans, which means understanding what humans somewhere want. It means avoiding reliance on algorithms and ensuring that you rely on the team's own human intuition. It's the human intuition that will bridge the gap between what the models can do by themselves and what people actually want.

Succeeding at the human side demands more human interaction instead of more time spent developing models.

The human side of data science is often also the hidden side. By understanding the way that your users will engage with what you have created, you will ensure that they will appreciate what you have created to its fullest potential. You will also make it far more likely that you will be asked back to help on more projects.

As long as your team remembers that being asked back isn't guaranteed, but is dependent on how well they have convinced your users of the value of your work, your team will go far in being accepted as being on the side of their users and being lauded for their achievements.

As long as you remember how the team's performance shapes how people value what you have done on your behalf, your reputation will grow and you will continue to enjoy increasingly exciting opportunities over the course of your career.

Data science means many things to many people. Regardless of where you come from, I hope—and believe—you can apply something from this book to making sure the data science products built in your organization succeed in making your users' lives easier and promote data science as a great tool to solve problems in many contexts.

Index

A

B

C

D

© Robert de Graaf 2019
R. de Graaf, *Managing Your Data Science Projects*,
https://doi.org/10.1007/978-1-4842-4907-9